Ground and Surface
Water Hydrology

Ground and Surface Water Hydrology

Edited by
Max Guthrie

⊟ Larsen & Keller
www.larsen-keller.com

Ground and Surface Water Hydrology
Edited by Max Guthrie
ISBN: 978-1-63549-694-9 (Hardback)

▤ Larsen & Keller

Published by Larsen and Keller Education,
5 Penn Plaza,
19th Floor,
New York, NY 10001, USA

Cataloging-in-Publication Data

Ground and surface water hydrology / edited by Max Guthrie.
    p. cm.
Includes bibliographical references and index.
ISBN 978-1-63549-694-9
1. Hydrology. 2. Groundwater. I. Guthrie, Max.
GB653 .G76 2018
551.48--dc23

For more information regarding Larsen and Keller Education and its products, please visit the publisher's website www.larsen-keller.com

Table of Contents

Preface

Ground water hydrology is the process of examining, analyzing and studying the movements, distribution and properties of water present on the surface of earth and in aquifers. Some of the key aspects studied under this field are porosity, hydraulic conductivity, hydraulic head, specific storage, water content, hydrodynamic dispersion, retardation by absorption, molecular diffusion, etc. This book is a compilation of chapters that discuss the most vital concepts in the field of ground water hydrology. It outlines the processes and applications of this subject in detail. Some of the diverse topics covered in it address the varied branches that fall under this category. Through this textbook, we attempt to further enlighten the readers about the new concepts in this field.

To facilitate a deeper understanding of the contents of this book a short introduction of every chapter is written below:

Chapter 1- Water recedes into the ground after rainfall through infiltration and is accumulated under the soil strata. This water is called groundwater. The science of distribution, occurrence, and movement of water below the Earth's surface is called groundwater hydrology. Other concepts discussed in this section are of aquifers, water cycle, groundwater model, etc. This chapter on groundwater hydrology offers an insightful focus, keeping in mind the complex subject matter.

Chapter 2- The movement of groundwater can be understood in the context of Darcy's law. It is an equation that describes the flow of water through a porous media. This permeability is normally denoted in Darcy units. The law is mainly used to estimate water flow in aquifers. This chapter elucidates the crucial theories and principles of groundwater flow.

Chapter 3- Water equilibrium can be achieved in confined and unconfined aquifer. If the pumping of water into a well is constant, the simultaneous refilling of the well causes water flow to be radially symmetric. The chapter closely examines the key concepts of water equilibrium to provide an extensive understanding of the subject.

Chapter 4- It is necessary to preserve aquifers. Here, groundwater management takes importance. It analyzes and predicts the condition of aquifers and aims to achieve a sustainable approach towards use of water. A few of its objectives are that it maximizes water withdrawal; it minimizes the energy required to extract and distribute water. This chapter has been carefully written to provide an easy understanding of the varied facets of groundwater preservation.

Chapter 5- Groundwater can be contaminated through various sources such as radioactive waste disposal sites, septic tanks and cesspools, mine wastes, landfill leaching, animal burials, pesticide and fertilizers applied to crop field, etc. The contaminating solute in groundwater can spread through dispersion and advection. The aspects elucidated in this section are of vital importance, and provide a better understanding of pollutant flow.

I would like to share the credit of this book with my editorial team who worked tirelessly on this book. I owe the completion of this book to the never-ending support of my family, who supported me throughout the project.

Editor

Fundamentals of Groundwater Hydrology

Water recedes into the ground after rainfall through infiltration and is accumulated under the soil strata. This water is called groundwater. The science of distribution, occurrence, and movement of water below the Earth's surface is called groundwater hydrology. Other concepts discussed in this section are of aquifers, water cycle, groundwater model, etc. This chapter on groundwater hydrology offers an insightful focus, keeping in mind the complex subject matter.

Groundwater Hydrology

A major component of precipitation that falls on the earth surface eventually enters into the ground by the process of infiltration. The infiltrated water is stored in the pores of the underground soil strata. The water which is stored in the pores of the soil strata is known as groundwater. Therefore, the groundwater may be defined as all the water present below the earth surface and the groundwater hydrology is defined as the science of occurrence, distribution and movement of water below the earth surface. In this section of hydrology, we generally deal with the water that is stored in the voids of the soil below the earth surface and also their interaction with the water that are present above the earth surface. When all the pores of a soil matrix are filled with water, we call that the soil is in the state of saturation. We used the term *porosity* to quantify the amount of voids space available in a soil matrix. Porosity is defined as the ratio of volume of voids to the total volume of the soil matrix. The porosity is expressed as,

$$\eta = \frac{V_v}{V_T}$$

Where V_v is the volume of void and V_T is the total volume of soil solid.

Water entered into the earth surface also moves from one place to another through the pores of the underground strata. This is known as subsurface flow. The subsurface flow is three dimensional and can be estimated using Darcy's law. The subsurface water also comes out to the earth surface as spring, river base flow, *etc* . and also goes back to the atmosphere by the process of evapo- transpiration. Thus this is a continuous process of recycling of water from the atmosphere down to the soil below the earth surface and back to the atmosphere again. This cycle below is called hydro-geological cycle. In this process of recycling, the water molecules spent some time under the earth surface. The average length of time spent by the water molecules under the earth surface is known as residence time of groundwater. The residence time can be calculated as,

$$t_r = \frac{V_{gr}}{q_{av}}$$

Where, t_r is the residence time for groundwater, V_{gr} is the volume of groundwater and q_{av} is the inflow or outflow at steady rate.

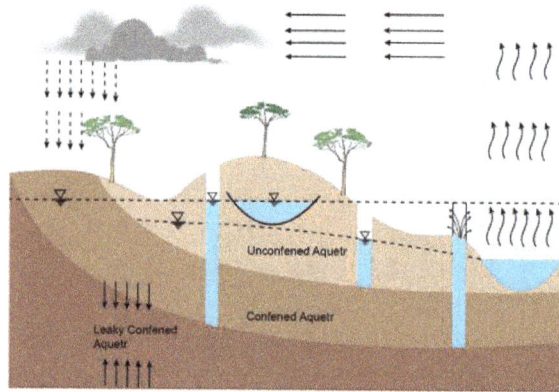

Hydro-geological cycle

Classification of Ground Water

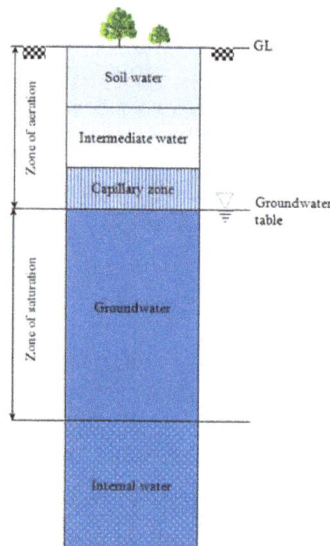

There are various classifications of groundwater given by different researchers. However, as per the most popular classification given by Meinzer (1923), the groundwater has been divided mainly in two groups: interstitial water and internal water. The interstitial water is again subdivided into two divisions. They are vadose water present in the zone of aeration and groundwater present in the zone of saturation. The vadose water is further subdivided into three zones, *i.e.*, soil water zone, intermediate zone and capillary zone. Above figure shows the classification of groundwater. The soil water zone is adjacent to the ground surface. The intermediate zone is between the lower edge of the soil water zone and the upper edge of the capillary zone. The capillary zone extends from the bottom edge of the intermediate zone to the upper edge of the saturated zone. The thickness of the capillary zone depends on the properties of the soil and also on the homogeneity of the soil. The depth of capillary zone is varying from few centimeters to few meters. In capillary zone, all the pores are field up with water. However, we cannot draw water by inserting a well up to that depth. This is because of the negative pressure developed at this zone due to surface tension effect. Groundwater zone starts from the bottom edge of the capillary zone. In this zone, all the pores of the soil matrix are filled with water. This zone is also known as zone of saturation. The top surface of the zone of saturation or groundwater is known as phreatic surface. This phreatic surface is also known as water table. Figure above shows Classification of groundwater.

The degree of saturation for the soil below the water table is equal to 1, *i.e.* the soil is fully saturated. As a groundwater hydrologist, we are primarily interested for the water below the groundwater table, *i.e.* the water available in the zone of saturation . For the soil above the water table, the degree of saturation of the soil is varying between 0 and 1 . However, the degree of saturation will never be 0 due of the presence of hygroscopic water. The hygroscopic water is the water that held tightly on the surface of the soil colloidal particle. Hygroscopic water can be removed from the soil by oven drying. Figure below shows the moisture distribution in soil column.

Approximation of Groundwater Table

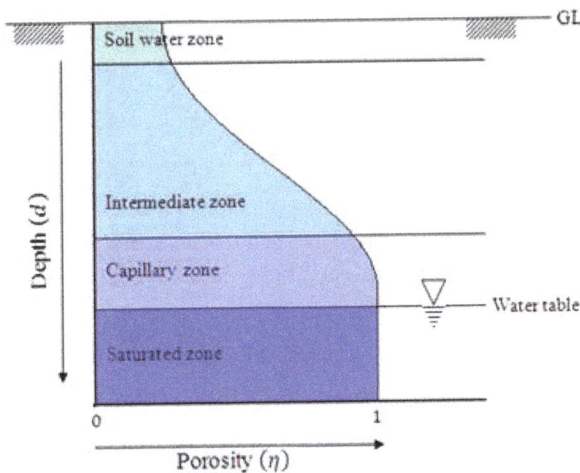

Moisture distribution in a soil column

The water table acts as a boundary between saturated zone and unsaturated zone. The soil matrix is fully saturated below the water table. At the same time, the soil just above the water table is also saturated due to the capillary effect. The depth of capillary rise may be from few centimeters to few meters. A suggested by Silin Bekchurin (1958), capillary rise may be around 2-5 cm in case of course sand, may be around 12-35 cm in case of sand, around 35-70 cm in case of fine sand, around 70-150 cm in case of silt and around 2-4 m and more in case of clay soil. Figure below shows the actual and approximate distribution of the moisture content. The actual distribution can be approximate by a step function which is necessary to approximate the elevation of the groundwater table. The step defines the depth of the capillary rise, h_c.

It can be assumed that up to the distance of h_c, above the phreatic surface, the aquifer is fully saturated. The aquifer above h_c line is completely dry, i.e. no moisture is present. The upper end of the capillary fringe may be taken as the groundwater table. However, when depth of capillary fringe, h_c is much smaller than the thickness of the aquifer below the water table, the capillary fringe may be neglected in solving real world problems. The depth of the capillary fringe can be approximated as (Mavis and Tsui 1939)

$$h_c = \frac{2.2}{dm}\left(\frac{1-n}{n}\right)^{3/2}$$

where d_m is the mean diameter of the soil grain, n is the porosity.

Polubarinova - Kochina (1952, 1962) approximated the capillary fringe as

$$h_c = \frac{0.45}{d_{10}}\left(\frac{1-n}{n}\right)$$

where d_{10} is the partical size at which 10% of the total partical is finer than that size.

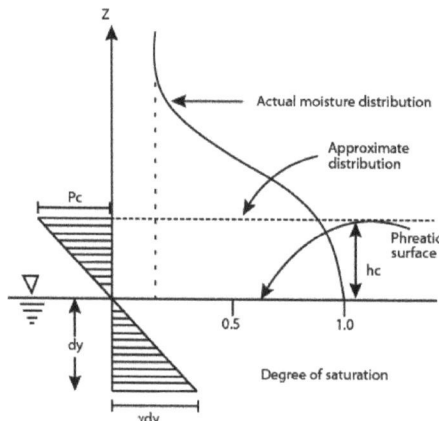

Approximation of groundwater table

Aquifer

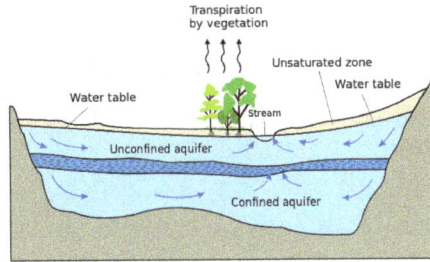

Typical aquifer cross-section

An aquifer is an underground layer of water-bearing permeable rock, rock fractures or unconsolidated materials (gravel, sand, or silt) from which groundwater can be extracted using a water well. The study of water flow in aquifers and the characterization of aquifers is called hydrogeology. Related terms include aquitard, which is a bed of low permeability along an aquifer, and aquiclude (or *aquifuge*), which is a solid, impermeable area underlying or overlying an aquifer. If the impermeable area overlies the aquifer, pressure could cause it to become a confined aquifer.

Depth

Aquifers may occur at various depths. Those closer to the surface are not only more likely to be used for water supply and irrigation, but are also more likely to be topped up by the local rainfall. Many desert areas have limestone hills or mountains within them or close to them that can be exploited as groundwater resources. Part of the Atlas Mountains in North Africa, the Lebanon and Anti-Lebanon ranges between Syria and Lebanon, the Jebel Akhdar (Oman) in Oman, parts of the Sierra Nevada and neighboring ranges in the United States' Southwest, have shallow aquifers that are exploited for their water. Overexploitation can lead to the exceeding of the practical sustained yield; i.e., more water is taken out than can be replenished. Along the coastlines of certain countries, such as Libya and Israel, increased water usage associated with population growth has caused a lowering of the water table and the subsequent contamination of the groundwater with saltwater from the sea.

The beach provides a model to help visualize an aquifer. If a hole is dug into the sand, very wet or saturated sand will be located at a shallow depth. This hole is a crude well, the wet sand represents an aquifer, and the level to which the water rises in this hole represents the water table.

In 2013 large freshwater aquifers were discovered under continental shelves off Australia, China, North America and South Africa. They contain an estimated half a million cubic kilometers of "low salinity" water that could be economically processed into potable water. The reserves formed when ocean levels were lower and rainwater made its way into the ground in land areas that were not submerged until the ice age ended 20,000 years ago. The volume is estimated to be 100x the amount of water extracted from other aquifers since 1900.

Classification

The above diagram (Typical aquifer cross-section) indicates typical flow directions in a cross-sectional view of a simple confined or unconfined aquifer system. The system shows two aquifers with one aquitard (a confining or impermeable layer) between them, surrounded by the bedrock *aquiclude*, which is in contact with a gaining stream (typical in humid regions). The water table and unsaturated zone are also illustrated. An *aquitard* is a zone within the earth that restricts the flow of groundwater from one aquifer to another. An aquitard can sometimes, if completely impermeable, be called an *aquiclude* or *aquifuge*. Aquitards are composed of layers of either clay or non-porous rock with low hydraulic conductivity.

Saturated Versus Unsaturated

Groundwater can be found at nearly every point in the Earth's shallow subsurface to some degree, although aquifers do not necessarily contain fresh water. The Earth's crust can be divided into two regions: the *saturated zone* or *phreatic zone* (e.g., aquifers, aquitards, etc.), where all available spaces are filled with water, and the *unsaturated zone* (also called the vadose zone), where there are still pockets of air that contain some water, but can be filled with more water.

Saturated means the pressure head of the water is greater than atmospheric pressure (it has a gauge pressure > 0). The definition of the water table is the surface where the pressure head is equal to atmospheric pressure (where gauge pressure = 0).

Unsaturated conditions occur above the water table where the pressure head is negative (absolute pressure can never be negative, but gauge pressure can) and the water that incompletely fills the pores of the aquifer material is under suction. The water content in the unsaturated zone is held in place by surface adhesive forces and it rises above the water table (the zero-gauge-pressure isobar) by capillary action to saturate a small zone above the phreatic surface (the capillary fringe) at less than atmospheric pressure. This is termed tension saturation and is not the same as saturation on a water-content basis. Water content in a capillary fringe decreases with increasing distance from the phreatic surface. The capillary head depends on soil pore size. In sandy soils with larger pores, the head will be less than in clay soils with very small pores. The normal capillary rise in a clayey soil is less than 1.80 m (six feet) but can range between 0.3 and 10 m (one and 30 ft).

The capillary rise of water in a small-diameter tube involves the same physical process. The water table is the level to which water will rise in a large-diameter pipe (e.g., a well) that goes down into the aquifer and is open to the atmosphere.

Aquifers Versus Aquitards

Aquifers are typically saturated regions of the subsurface that produce an economically feasible quantity of water to a well or spring (e.g., sand and gravel or fractured bedrock often make good aquifer materials).

An aquitard is a zone within the earth that restricts the flow of groundwater from one aquifer to another. A completely impermeable aquitard is called an aquiclude or aquifuge. Aquitards comprise layers of either clay or non-porous rock with low hydraulic conductivity.

In mountainous areas (or near rivers in mountainous areas), the main aquifers are typically unconsolidated alluvium, composed of mostly horizontal layers of materials deposited by water processes (rivers and streams), which in cross-section (looking at a two-dimensional slice of the aquifer) appear to be layers of alternating coarse and fine materials. Coarse materials, because of the high energy needed to move them, tend to be found nearer the source (mountain fronts or rivers), whereas the fine-grained material will make it farther from the source (to the flatter parts of the basin or overbank areas - sometimes called the pressure area). Since there are less fine-grained deposits near the source, this is a place where aquifers are often unconfined (sometimes called the forebay area), or in hydraulic communication with the land surface.

Confined Versus Unconfined

There are two end members in the spectrum of types of aquifers; *confined* and *unconfined* (with semi-confined being in between). Unconfined aquifers are sometimes also called *water table* or *phreatic* aquifers, because their upper boundary is the water table or phreatic surface. Typically (but not always) the shallowest aquifer at a given location is unconfined, meaning it does not have a confining layer (an aquitard or aquiclude) between it and the surface. The term "perched" refers to ground water accumulating above a low-permeability unit or strata, such as a clay layer. This term is generally used to refer to a small local area of ground water that occurs at an elevation higher than a regionally extensive aquifer. The difference between perched and unconfined aquifers is their size (perched is smaller). Confined aquifers are aquifers that are overlain by a confining layer, often made up of clay. The confining layer might offer some protection from surface contamination.

If the distinction between confined and unconfined is not clear geologically (i.e., if it is not known if a clear confining layer exists, or if the geology is more complex, e.g., a fractured bedrock aquifer), the value of storativity returned from an aquifer test can be used to determine it (although aquifer tests in unconfined aquifers should be interpreted differently than confined ones). Confined aquifers have very low storativity values (much less than 0.01, and as little as 10^{-5}), which means that the aquifer is storing water using the mechanisms of aquifer matrix expansion and the compressibility of water, which typically are both quite small quantities. Unconfined aquifers have storativities (typically then called specific yield) greater than 0.01 (1% of bulk volume); they release water from storage by the mechanism of actually draining the pores of the aquifer, releasing relatively large amounts of water (up to the drainable porosity of the aquifer material, or the minimum volumetric water content).

Isotropic Versus Anisotropic

In isotropic aquifers or aquifer layers the hydraulic conductivity (K) is equal for flow in all directions, while in anisotropic conditions it differs, notably in horizontal (Kh) and vertical (Kv) sense.

Semi-confined aquifers with one or more aquitards work as an anisotropic system, even when the separate layers are isotropic, because the compound Kh and Kv values are different.

When calculating flow to drains or flow to wells in an aquifer, the anisotropy is to be taken into account lest the resulting design of the drainage system may be faulty.

Groundwater in Rock Formations

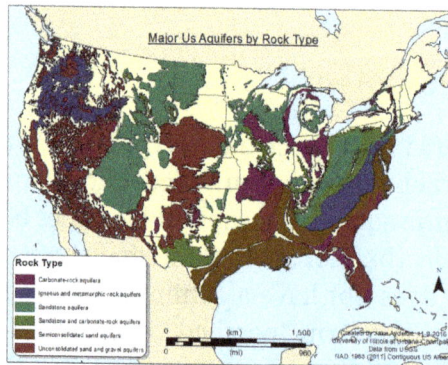

Map of major US aquifers by rock type

Groundwater may exist in *underground rivers* (e.g., caves where water flows freely underground). This may occur in eroded limestone areas known as karst topography, which make up only a small percentage of Earth's area. More usual is that the pore spaces of rocks in the subsurface are simply saturated with water — like a kitchen sponge — which can be pumped out for agricultural, industrial, or municipal uses.

If a rock unit of low porosity is highly fractured, it can also make a good aquifer (via fissure flow), provided the rock has a hydraulic conductivity sufficient to facilitate movement of water. Porosity is important, but, *alone*, it does not determine a rock's ability to act as an aquifer. Areas of the Deccan Traps (a basaltic lava) in west central India are good examples of rock formations with high porosity but low permeability, which makes them poor aquifers. Similarly, the micro-porous (Upper Cretaceous) Chalk of south east England, although having a reasonably high porosity, has a low grain-to-grain permeability, with its good water-yielding characteristics mostly due to micro-fracturing and fissuring.

Human Dependence on Groundwater

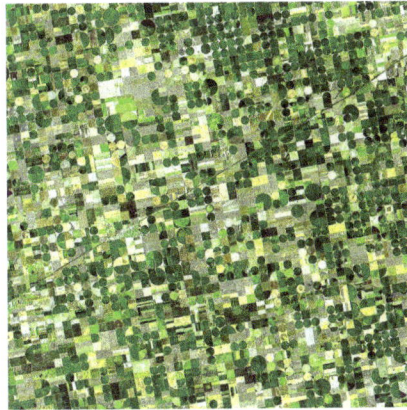

Center-pivot irrigated fields in Kansas covering hundreds of
square miles watered by the Ogallala Aquifer

Most land areas on Earth have some form of aquifer underlying them, sometimes at significant depths. In some cases, these aquifers are rapidly being depleted by the human population.

Fresh-water aquifers, especially those with limited recharge by snow or rain, also known as meteoric water, can be over-exploited and depending on the local hydrogeology, may draw in non-potable water or saltwater intrusion from hydraulically connected aquifers or surface water bodies. This can be a serious problem, especially in coastal areas and other areas where aquifer pumping is excessive. In some areas, the ground water can become contaminated by arsenic and other mineral poisons.

Aquifers are critically important in human habitation and agriculture. Deep aquifers in arid areas have long been water sources for irrigation. Many villages and even large cities draw their water supply from wells in aquifers.

Municipal, irrigation, and industrial water supplies are provided through large wells. Multiple wells for one water supply source are termed "wellfields", which may withdraw water from confined or unconfined aquifers. Using ground water from deep, confined aquifers provides more protection from surface water contamination. Some wells, termed "collector wells," are specifically designed to induce infiltration of surface (usually river) water.

Aquifers that provide sustainable fresh groundwater to urban areas and for agricultural irrigation are typically close to the ground surface (within a couple of hundred metres) and have some recharge by fresh water. This recharge is typically from rivers or meteoric water (precipitation) that percolates into the aquifer through overlying unsaturated materials.

Occasionally, sedimentary or "fossil" aquifers are used to provide irrigation and drinking water to urban areas. In Libya, for example, Muammar Gaddafi's Great Manmade River project has pumped large amounts of groundwater from aquifers beneath the Sahara to populous areas near the coast. Though this has saved Libya money over the alternative, desalination, the aquifers are likely to run dry in 60 to 100 years. Aquifer depletion has been cited as one of the causes of the food price rises of 2011.

Subsidence

In unconsolidated aquifers, groundwater is produced from pore spaces between particles of gravel, sand, and silt. If the aquifer is confined by low-permeability layers, the reduced water pressure in the sand and gravel causes slow drainage of water from the adjoining confining layers. If these confining layers are composed of compressible silt or clay, the loss of water to the aquifer reduces the water pressure in the confining layer, causing it to compress from the weight of overlying geologic materials. In severe cases, this compression can be observed on the ground surface as subsidence. Unfortunately, much of the subsidence from groundwater extraction is permanent (elastic rebound is small). Thus, the subsidence is not only permanent, but the compressed aquifer has a permanently reduced capacity to hold water.

Saltwater Intrusion

Aquifers near the coast have a lens of freshwater near the surface and denser seawater under freshwater. Seawater penetrates the aquifer diffusing in from the ocean and is denser than freshwater. For porous (i.e., sandy) aquifers near the coast, the thickness of freshwater atop saltwater is about 40 feet (12 m) for every 1 ft (0.30 m) of freshwater head above sea level. This relationship is called the Ghyben-Herzberg equation. If too much ground water is pumped near the coast, salt-water may intrude into freshwater aquifers causing contamination of potable freshwater supplies. Many coastal aquifers, such as the Biscayne Aquifer near Miami and the New Jersey Coastal Plain aquifer, have problems with saltwater intrusion as a result of overpumping and sea level rise.

Salination

Aquifers in surface irrigated areas in semi-arid zones with reuse of the unavoidable irrigation water losses percolating down into the underground by supplemental irrigation from wells run the risk of salination.

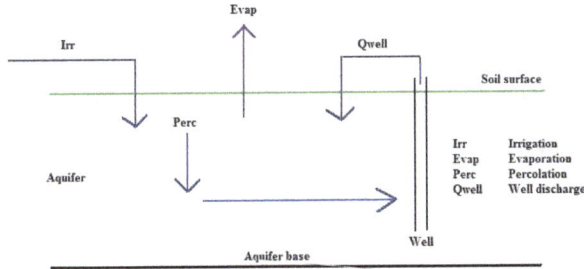

Diagram of a water balance of the aquifer

Surface irrigation water normally contains salts in the order of 0.5 g/l or more and the annual irrigation requirement is in the order of 10000 m³/ha or more so the annual import of salt is in the order of 5000 kg/ha or more.

Under the influence of continuous evaporation, the salt concentration of the aquifer water may increase continually and eventually cause an environmental problem.

For salinity control in such a case, annually an amount of drainage water is to be discharged from the aquifer by means of a subsurface drainage system and disposed of through a safe outlet. The drainage system may be *horizontal* (i.e. using pipes, tile drains or ditches) or *vertical* (drainage by wells). To estimate the drainage requirement, the use of a groundwater model with an agro-hydro-salinity component may be instrumental, e.g. SahysMod.

Examples

The Great Artesian Basin situated in Australia is arguably the largest groundwater aquifer in the world (over 1.7 million km²). It plays a large part in water supplies for Queensland and remote parts of South Australia.

The Guarani Aquifer, located beneath the surface of Argentina, Brazil, Paraguay, and Uruguay, is one of the world's largest aquifer systems and is an important source of fresh water. Named after the Guarani people, it covers 1,200,000 km², with a volume of about 40,000 km³, a thickness of between 50 m and 800 m and a maximum depth of about 1,800 m.

Aquifer depletion is a problem in some areas, and is especially critical in northern Africa, for example the Great Manmade River project of Libya. However, new methods of groundwater management such as artificial recharge and injection of surface waters during seasonal wet periods has extended the life of many freshwater aquifers, especially in the United States.

The Ogallala Aquifer of the central United States is one of the world's great aquifers, but in places it is being rapidly depleted by growing municipal use, and continuing agricultural use. This huge aquifer, which underlies portions of eight states, contains primarily fossil water from the time of the last glaciation. Annual recharge, in the more arid parts of the aquifer, is estimated to total only about 10 percent of annual withdrawals. According to a 2013 report by research hydrologist Leonard F. Konikow at the United States Geological Survey (USGS), the depletion between 2001–2008, inclusive, is about 32 percent of the cumulative depletion during the entire 20th century (Konikow 2013:22)." In the United States, the biggest users of water from aquifers include agricultural irrigation and oil and coal extraction. "Cumulative total groundwater depletion in the United States accelerated in the late 1940s and continued at an almost steady linear rate through the end of the century. In addition to widely recognized environmental consequences, groundwater depletion also adversely impacts the long-term sustainability of groundwater supplies to help meet the Nation's water needs."

An example of a significant and sustainable carbonate aquifer is the Edwards Aquifer in central Texas. This carbonate aquifer has historically been providing high quality water for nearly 2 million people, and even today, is full because of tremendous recharge from a number of area streams, rivers and lakes. The primary risk to this resource is human development over the recharge areas.

Discontinuous sand bodies at the base of the McMurray Formation in the Athabasca Oil Sands region of northeastern Alberta, Canada, are commonly referred to as the Basal Water Sand (BWS) aquifers. Saturated with water, they are confined beneath impermeable bitumen-saturated sands that are exploited to recover bitumen for synthetic crude oil production. Where they are deep-lying and recharge occurs from underlying Devonian formations they are saline, and where they are shallow and recharged by meteoric water they are non-saline. The BWS typically pose problems for the recovery of bitumen, whether by open-pit mining or by *in situ* methods such as steam-assisted gravity drainage (SAGD), and in some areas they are targets for waste-water injection.

Water Cycle

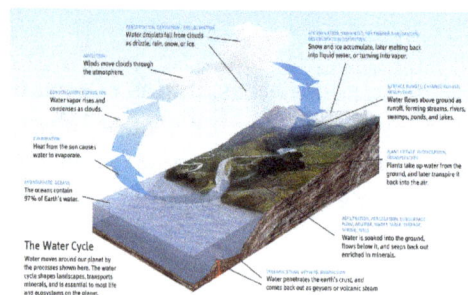

Diagram of the Water Cycle

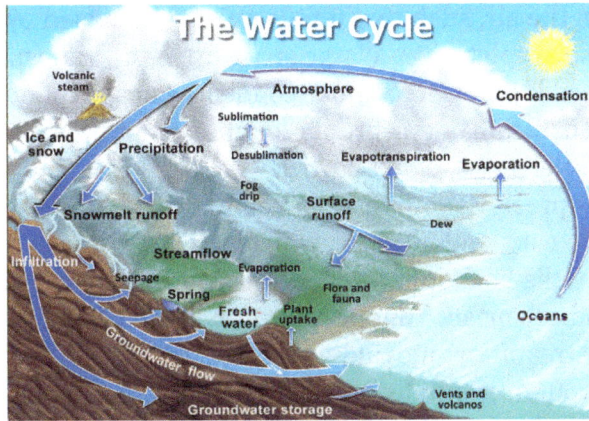

The water cycle

The water cycle, also known as the hydrological cycle or the hydrologic cycle, describes the continuous movement of water on, above and below the surface of the Earth. The mass of water on Earth remains fairly constant over time but the partitioning of the water into the major reservoirs of ice, fresh water, saline water and atmospheric water is variable depending on a wide range of climatic variables. The water moves from one reservoir to another, such as from river to ocean, or from the ocean to the atmosphere, by the physical processes of evaporation, condensation, precipitation, infiltration, surface runoff, and subsurface flow. In doing so, the water goes through different forms: liquid, solid (ice) and vapor.

The water cycle involves the exchange of energy, which leads to temperature changes. For instance, when water evaporates, it takes up energy from its surroundings and cools the environment. When it condenses, it releases energy and warms the environment. These heat exchanges influence climate.

The evaporative phase of the cycle purifies water which then replenishes the land with freshwater. The flow of liquid water and ice transports minerals across the globe. It is also involved in reshaping the geological features of the Earth, through processes including erosion and sedimentation. The water cycle is also essential for the maintenance of most life and ecosystems on the planet.

Description

The sun, which drives the water cycle, heats water in oceans and seas. Water evaporates as water vapor into the air. Ice and snow can sublimate directly into water vapour. Evapotranspiration is water transpired from plants and evaporated from the soil. The water vapour molecule H_2O has less density compared to the major components of the atmosphere, nitrogen and oxen, N_2 and O_2. Due to the significant difference in molecular mass, water vapor in gas form gains height in open air as a result of buoyancy. However, as altitude increases, air pressure decreases and the temperature drops. The lowered temperature causes water vapour to condense into a tiny liquid water droplet which is heavier than the air, such that it falls unless supported by an updraft. A huge concentration of these droplets over a large space up in the atmosphere become visible as cloud. Fog is formed if the water vapour condenses near ground level, as a result of moist air and cool air collision or an abrupt reduction in air pressure. Air currents move water vapour around the globe, cloud particles collide, grow, and fall out of the upper atmospheric layers as precipitation. Some precipitation falls as snow or hail, sleet, and can accumulate as ice caps and glaciers, which can store frozen water for thousands of years. Most water falls back into the oceans or onto land as rain, where the water flows over the ground as surface runoff. A portion of runoff enters rivers in valleys in the landscape, with streamflow moving water towards the oceans. Runoff and water emerging from the ground (groundwater) may be stored as freshwater in lakes. Not all runoff flows into rivers, much of it soaks into the ground as infiltration. Some water infiltrates deep into the ground and replenishes aquifers, which can store freshwater for long periods of time. Some infiltration stays close to the land surface and can seep back into surface-water bodies (and the ocean) as groundwater discharge. Some groundwater finds openings in the land surface and comes out as freshwater springs. In river valleys and floodplains, there is often continuous water exchange between surface water and ground water in the hyporheic zone. Over time, the water returns to the ocean, to continue the water cycle.

Processes

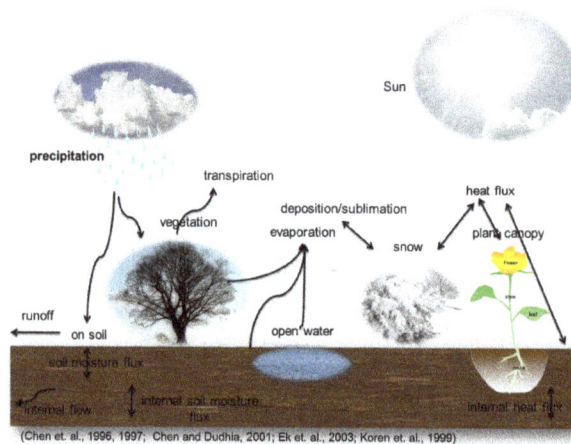

(Chen et. al., 1996, 1997; Chen and Dudhia, 2001; Ek et. al., 2003; Koren et. al., 1999)

Many different processes lead to movements and phase changes in water

Precipitation

Condensed water vapor that falls to the Earth's surface. Most precipitation occurs as rain, but also includes snow, hail, fog drip, graupel, and sleet. Approximately 505,000 km³ (121,000 cu mi) of water falls as precipitation each year, 398,000 km³ (95,000 cu mi) of it over the oceans. The rain on land contains 107,000 km³ (26,000 cu mi) of water per year and a snowing only 1,000 km³ (240 cu mi). 78% of global precipitation occurs over the ocean.

Canopy interception

The precipitation that is intercepted by plant foliage eventually evaporates back to the atmosphere rather than falling to the ground.

Snowmelt

The runoff produced by melting snow.

Runoff

The variety of ways by which water moves across the land. This includes both surface runoff and channel runoff. As it flows, the water may seep into the ground, evaporate into the air, become stored in lakes or reservoirs, or be extracted for agricultural or other human uses.

Infiltration

The flow of water from the ground surface into the ground. Once infiltrated, the water becomes soil moisture or groundwater. A recent global study using water stable isotopes, however, shows that not all soil moisture is equally available for groundwater recharge or for plant transpiration.

Subsurface flow

The flow of water underground, in the vadose zone and aquifers. Subsurface water may return to the surface (e.g. as a spring or by being pumped) or eventually seep into the oceans. Water returns to the land surface at lower elevation than where it infiltrated, under the force of gravity or gravity induced pressures. Groundwater tends to move slowly and is replenished slowly, so it can remain in aquifers for thousands of years.

Evaporation

The transformation of water from liquid to gas phases as it moves from the ground or bodies of water into the overlying atmosphere. The source of energy for evaporation is primarily solar radiation. Evaporation often implicitly includes transpiration from plants, though together they are specifically referred to as evapotranspiration. Total annual evapotranspiration amounts to approximately 505,000 km³ (121,000 cu mi) of water, 434,000 km³ (104,000 cu mi) of which evaporates from the oceans. 86% of global evaporation occurs over the ocean.

Sublimation

The state change directly from solid water (snow or ice) to water vapor.

Deposition

This refers to changing of water vapor directly to ice.

Advection

> The movement of water — in solid, liquid, or vapor states — through the atmosphere. Without advection, water that evaporated over the oceans could not precipitate over land.

Condensation

> The transformation of water vapor to liquid water droplets in the air, creating clouds and fog.

Transpiration

> The release of water vapor from plants and soil into the air. Water vapor is a gas that cannot be seen.

Percolation

> Water flows vertically through the soil and rocks under the influence of gravity

Plate tectonics

> Water enters the mantle via subduction of oceanic crust. Water returns to the surface via volcanism.

Water cycle thus involves many of the intermediate processes.

Residence Times

Average reservoir residence times	
Reservoir	**Average residence time**
Antarctica	20,000 years
Oceans	3,200 years
Glaciers	20 to 100 years
Seasonal snow cover	2 to 6 months
Soil moisture	1 to 2 months
Groundwater: shallow	100 to 200 years
Groundwater: deep	10,000 years
Lakes	50 to 100 years
Rivers	2 to 6 months
Atmosphere	9 days

The *residence time* of a reservoir within the hydrologic cycle is the average time a water molecule will spend in that reservoir. It is a measure of the average age of the water in that reservoir.

Groundwater can spend over 10,000 years beneath Earth's surface before leaving. Particularly old groundwater is called fossil water. Water stored in the soil remains there very briefly, because it is spread thinly across the Earth, and is readily lost by evaporation, transpiration, stream flow, or groundwater recharge. After evaporating, the residence time in the atmosphere is about 9 days before condensing and falling to the Earth as precipitation.

The major ice sheets - Antarctica and Greenland - store ice for very long periods. Ice from Antarctica has been reliably dated to 800,000 years before present, though the average residence time is shorter.

In hydrology, residence times can be estimated in two ways. The more common method relies on the principle of conservation of mass and assumes the amount of water in a given reservoir is roughly constant. With this method, residence times are estimated by dividing the volume of the reservoir by the rate by which water either enters or exits the reservoir. Conceptually, this is equivalent to timing how long it would take the reservoir to become filled from empty if no water were to leave (or how long it would take the reservoir to empty from full if no water were to enter).

An alternative method to estimate residence times, which is gaining in popularity for dating groundwater, is the use of isotopic techniques. This is done in the subfield of isotope hydrology.

Changes Over Time

The water cycle describes the processes that drive the movement of water throughout the hydrosphere. However, much more water is "in storage" for long periods of time than is actually moving through the cycle. The storehouses for the vast majority of all water on Earth are the oceans. It is estimated that of the 332,500,000 mi³ (1,386,000,000 km³) of the world's water supply, about 321,000,000 mi³ (1,338,000,000 km³) is stored in oceans, or about 97%. It is also estimated that the oceans supply about 90% of the evaporated water that goes into the water cycle.

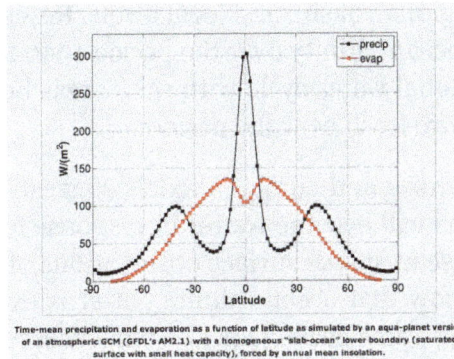

Time-mean precipitation and evaporation as a function of latitude as simulated by an aqua-planet version of an atmospheric GCM (GFDL's AM2.1) with a homogeneous "slab-ocean" lower boundary (saturated surface with small heat capacity), forced by annual mean insolation.

Time-mean precipitation and evaporation as a function of latitude as simulated by an aqua-planet version of an atmospheric GCM (GFDL's AM2.1) with a homogeneous "slab-ocean" lower boundary (saturated surface with small heat capacity), forced by annual mean insolation.

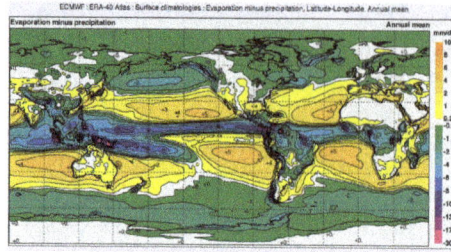

Global map of annual mean evaporation minus precipitation by latitude-longitude

During colder climatic periods more ice caps and glaciers form, and enough of the global water supply accumulates as ice to lessen the amounts in other parts of the water cycle. The reverse is true during warm periods. During the last ice age glaciers covered almost one-third of Earth's land mass, with the result being that the oceans were about 400 ft (122 m) lower than today. During the last global "warm spell," about 125,000 years ago, the seas were about 18 ft (5.5 m) higher than they are now. About three million years ago the oceans could have been up to 165 ft (50 m) higher.

The scientific consensus expressed in the 2007 Intergovernmental Panel on Climate Change (IPCC) Summary for Policymakers is for the water cycle to continue to intensify throughout the 21st century, though this does not mean that precipitation will increase in all regions. In subtropical land areas — places that are already relatively dry — precipitation is projected to decrease during the 21st century, increasing the probability of drought. The drying is projected to be strongest near the poleward margins of the subtropics (for example, the Mediterranean Basin, South Africa, southern Australia, and the Southwestern United States). Annual precipitation amounts are expected to increase in near-equatorial regions that tend to be wet in the present climate, and also at high latitudes. These large-scale patterns are present in nearly all of the climate model simulations conducted at several international research centers as part of the 4th Assessment of the IPCC. There is now ample evidence that increased hydrologic variability and change in climate has and will continue to have a profound impact on the water sector through the hydrologic cycle, water availability, water demand, and water allocation at the global, regional, basin, and local levels. Research published in 2012 in *Science* based on surface ocean salinity over the period 1950 to 2000 confirm this projection of an intensified global water cycle with salty areas becoming more saline and fresher areas becoming more fresh over the period:

Fundamental thermodynamics and climate models suggest that dry regions will become drier and wet regions will become wetter in response to warming. Efforts to detect this long-term response in sparse surface observations of rainfall and evaporation remain ambiguous. We show that ocean salinity patterns express an identifiable fingerprint of an intensifying water cycle. Our 50-year observed global surface salinity changes, combined with changes from global climate models, present robust evidence of an intensified global water cycle at a rate of 8 ± 5% per degree of surface warming. This rate is double the response projected by current-generation climate models and suggests that a substantial (16 to 24%) intensification of the global water cycle will occur in a future 2° to 3° warmer world.

An instrument carried by the SAC-D satellite launched in June, 2011 measures global sea surface salinity but data collection began only in June, 2011.

Glacial retreat is also an example of a changing water cycle, where the supply of water to glaciers from precipitation cannot keep up with the loss of water from melting and sublimation. Glacial retreat since 1850 has been extensive.

Human activities that alter the water cycle include:

- agriculture

- industry

- alteration of the chemical composition of the atmosphere

- construction of dams

- deforestation and afforestation

- removal of groundwater from wells

- water abstraction from rivers

- urbanization

Effects on Climate

The water cycle is powered from solar energy. 86% of the global evaporation occurs from the oceans, reducing their temperature by evaporative cooling. Without the cooling, the effect of evaporation on the greenhouse effect would lead to a much higher surface temperature of 67 °C (153 °F), and a warmer planet.

Aquifer drawdown or overdrafting and the pumping of fossil water increases the total amount of water in the hydrosphere, and has been postulated to be a contributor to sea-level rise.

Effects on Biogeochemical Cycling

While the water cycle is itself a biogeochemical cycle, flow of water over and beneath the Earth is a key component of the cycling of other biogeochemicals. Runoff is responsible for almost all of the transport of eroded sediment and phosphorus from land to waterbodies. The salinity of the oceans is derived from erosion and transport of dissolved salts from the land. Cultural eutrophication of lakes is primarily due to phosphorus, applied in excess to agricultural fields in fertilizers, and then transported overland and down rivers. Both runoff and groundwater flow play significant roles in transporting nitrogen from the land to waterbodies. The dead zone at the outlet of the Mississippi River is a consequence of nitrates from fertilizer being carried off agricultural fields and funnelled down the river system to the Gulf of Mexico. Runoff also plays a part in the carbon cycle, again through the transport of eroded rock and soil.

Slow Loss Over Geologic Time

The hydrodynamic wind within the upper portion of a planet's atmosphere allows light chemical elements such as Hydrogen to move up to the exobase, the lower limit of the exosphere, where the gases can then reach escape velocity, entering outer space without impacting other particles of gas. This type of gas loss from a planet into space is known as planetary wind. Planets with hot lower atmospheres could result in humid upper atmospheres that accelerate the loss of hydrogen.

History of Hydrologic Cycle Theory

Floating Land Mass

In ancient times, it was thought that the land mass floated on a body of water, and that most of the water in rivers has its origin under the earth. Examples of this belief can be found in the works of Homer (circa 800 BCE).

Source of Rain

In the ancient near east, Hebrew scholars observed that even though the rivers ran into the sea, the sea never became full (Ecclesiastes 1:7). Some scholars conclude that the water cycle was described completely during this time in this passage: "The wind goeth toward the south, and turneth about unto the north; it whirleth about continually, and the wind returneth again according to its circuits. All the rivers run into the sea, yet the sea is not full; unto the place from whence the rivers come, thither they return again" (Ecclesiastes 1:6-7, KJV). Scholars are not in agreement as to the date of Ecclesiastes, though most scholars point to a date during the time of Solomon, the son of David and Bathsheba, "three thousand years ago, there is some agreement that the time period is 962-922 BCE. Furthermore, it was also observed that when the clouds were full, they emptied rain on the earth (Ecclesiastes 11:3). In addition, during 793-740 BC a Hebrew prophet, Amos, stated that water comes from the sea and is poured out on the earth (Amos 5:8, 9:6).

Precipitation and Percolation

In the Adityahridayam (a devotional hymn to the Sun God) of Ramayana, a Hindu epic dated to the 4th century BC, it is mentioned in the 22nd verse that the Sun heats up water and sends it down as rain. By roughly 500 BCE, Greek scholars were speculating that much of the water in rivers can be attributed to rain. The origin of rain was also known by then. These scholars maintained the belief, however, that water rising up through the earth contributed a great deal to rivers. Examples of this thinking included Anaximander (570 BCE) (who also speculated about the evolution of land animals from fish) and Xenophanes of Colophon (530 BCE). Chinese scholars such as Chi Ni Tzu (320 BC) and Lu Shih Ch'un Ch'iu (239 BCE) had similar thoughts. The idea that the water cycle is a closed cycle can be found in the works of Anaxagoras of Clazomenae (460 BCE) and Diogenes of Apollonia (460 BCE). Both Plato (390 BCE) and Aristotle (350 BCE) speculated about percolation as part of the water cycle.

Precipitation Alone

In the Biblical Book of Job, dated between 7th and 2nd centuries BCE, there is a description of precipitation in the hydrologic cycle, "For he maketh small the drops of water: they pour down rain according to the vapour thereof; Which the clouds do drop and distil upon man abundantly" (Job 36:27-28, KJV). Also found in the book of Ecclesiastes "All the rivers flow into the sea, Yet the sea is not full. To the place where the rivers flow, There they flow again." (Ecclesiastes 1:7)

Up to the time of the Renaissance, it was thought that precipitation alone was insufficient to feed rivers, for a complete water cycle, and that underground water pushing upwards from the oceans were the main contributors to river water. Bartholomew of England held this view (1240 CE), as did Leonardo da Vinci (1500 CE) and Athanasius Kircher (1644 CE).

The first published thinker to assert that rainfall alone was sufficient for the maintenance of rivers was Bernard Palissy (1580 CE), who is often credited as the "discoverer" of the modern theory of the water cycle. Palissy's theories were not tested scientifically until 1674, in a study commonly attributed to Pierre Perrault. Even then, these beliefs were not accepted in mainstream science until the early nineteenth century.

Groundwater Model

Groundwater models are computer models of groundwater flow systems, and are used by hydrogeologists. Groundwater models are used to simulate and predict aquifer conditions.

Characteristics

An unambiguous definition of "groundwater model" is difficult to give, but there are many common characteristics.

A groundwater model may be a scale model or an electric model of a groundwater situation or aquifer. Groundwater models are used to represent the natural groundwater flow in the environment. Some groundwater models include (chemical) quality aspects of the groundwater. Such groundwater models try to predict the fate and movement of the chemical in natural, urban or hypothetical scenario.

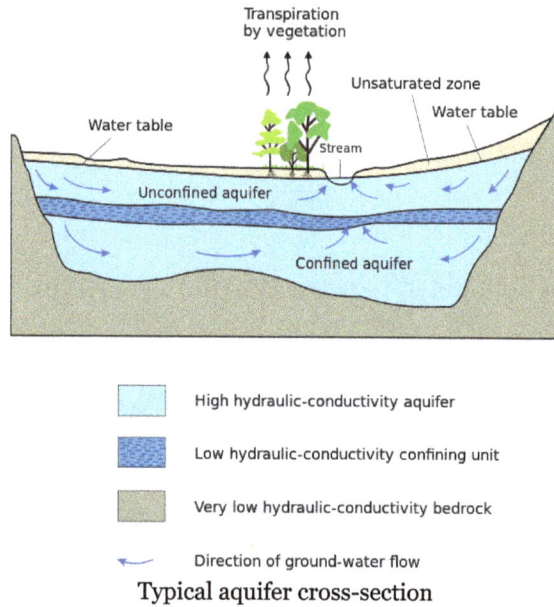

Typical aquifer cross-section

Groundwater models may be used to predict the effects of hydrological changes (like groundwater abstraction or irrigation developments) on the behavior of the aquifer and are often named groundwater simulation models. Also nowadays the groundwater models are used in various water management plans for urban areas.

As the computations in mathematical groundwater models are based on groundwater flow equations, which are differential equations that can often be solved only by approximate methods using a numerical analysis, these models are also called *mathematical, numerical, or computational groundwater models*.

The mathematical or the numerical models are usually based on the real physics the groundwater flow follows. These mathematical equations are solved using numerical codes such as MODFLOW, ParFlow, HydroGeoSphere, OpenGeoSys etc.

Inputs

For the calculations one needs inputs like:

- hydrological inputs,

- operational inputs,

- external conditions: initial and boundary conditions,

- (hydraulic) parameters.

The model may have chemical components like water salinity, soil salinity and other quality indicators of water and soil, for which inputs may also be needed.

Hydrological Inputs

The primary coupling between groundwater and hydrological inputs is the unsaturated zone or vadose zone. The soil acts to partition hydrological inputs such as rainfall or snowmelt into surface runoff, soil moisture, evapotranspiration and groundwater recharge. Flows through the unsaturated zone that couple surface water to soil moisture and groundwater can be upward or downward, depending upon the gradient of hydraulic head in the soil, can be modeled using the numerical solution of Richards' equation partial differential equation, or the ordinary differential equation Finite Water-Content method as validated for modeling groundwater and vadose zone interactions.

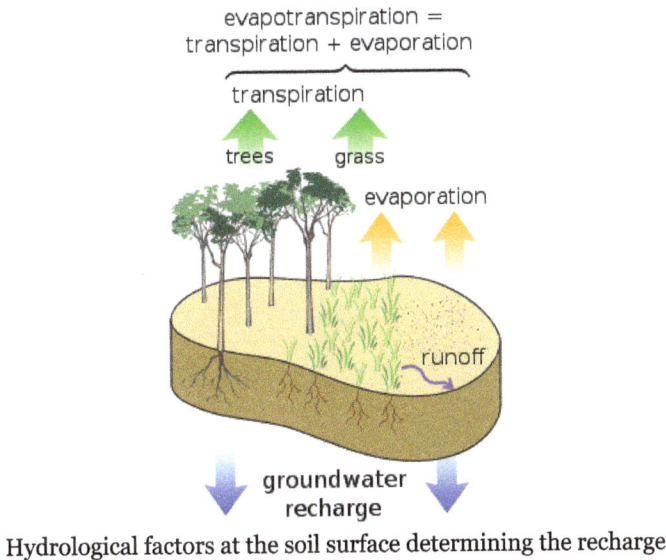

Hydrological factors at the soil surface determining the recharge

Operational Inputs

The operational inputs concern human interferences with the *water management* like irrigation, drainage, pumping from wells, watertable control, and the operation of retention or infiltration basins, which are often of an hydrological nature.

These inputs may also vary in time and space.

Many groundwater models are made for the purpose of assessing the effects hydraulic engineering measures.

Boundary and Initial Conditions

Boundary conditions can be related to levels of the water table, artesian pressures, and hydraulic head along the boundaries of the model on the one hand (the *head*

conditions), or to groundwater inflows and outflows along the boundaries of the model on the other hand (the *flow conditions*). This may also include quality aspects of the water like salinity.

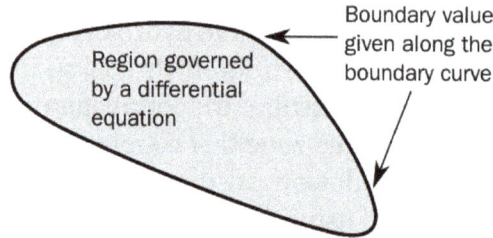

Boundary conditions

The *initial conditions* refer to initial values of elements that may increase or decrease in the course of the time *inside* the model domain and they cover largely the same phenomena as the boundary conditions do.

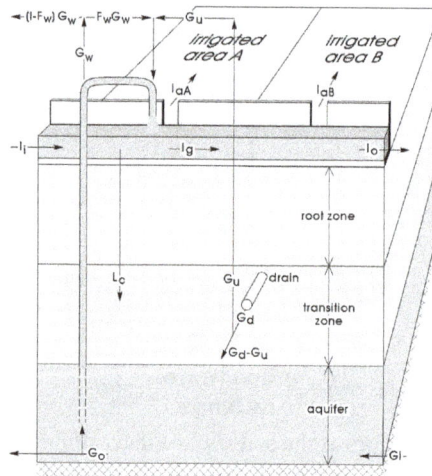

Example of parameters of an irrigation cum groundwater model

The initial and boundary conditions may vary from place to place. The boundary conditions may be kept either constant or be made variable in time.

Parameters

The parameters usually concern the geometry of and distances in the domain to be modelled and those physical properties of the aquifer that are more or less constant with time but that may be variable in space.

Important parameters are the topography, thicknesses of soil / rock layers and their horizontal/vertical hydraulic conductivity (permeability for water), aquifer transmissivity and resistance, aquifer porosity and storage coefficient, as well as the capillarity of the unsaturated zone.

Some parameters may be influenced by changes in the groundwater situation, like the thickness of a soil layer that may reduce when the water table drops and/the hydraulic pressure is reduced. This phenomenon is called subsidence. The thickness, in this case, is variable in time and not a parameter proper.

Applicability

The applicability of a groundwater model to a real situation depends on the accuracy of the input data and the parameters. Determination of these requires considerable study, like collection of hydrological data (rainfall, evapotranspiration, irrigation, drainage) and determination of the parameters mentioned before including pumping tests. As many parameters are quite variable in space, expert judgment is needed to arrive at representative values.

The models can also be used for the if-then analysis: if the value of a parameter is A, then what is the result, and if the value of the parameter is B instead, what is the influence? This analysis may be sufficient to obtain a rough impression of the groundwater behavior, but it can also serve to do a *sensitivity analysis* to answer the question: which factors have a great influence and which have less influence. With such information one may direct the efforts of investigation more to the influential factors.

When sufficient data have been assembled, it is possible to determine some of missing information by calibration. This implies that one assumes a range of values for the unknown or doubtful value of a certain parameter and one runs the model repeatedly while comparing results with known corresponding data. For example, if salinity figures of the groundwater are available and the value of hydraulic conductivity is uncertain, one assumes a range of conductivities and the selects that value of conductivity as "true" that yields salinity results close to the observed values, meaning that the groundwater flow as governed by the hydraulic conductivity is in agreemnent with the salinity conditions. This procedure is similar to the measurement of the flow in a river or canal by letting very saline water of a known salt concentration drip into the channel and measuring the resulting salt concentration downstream.

Dimensions

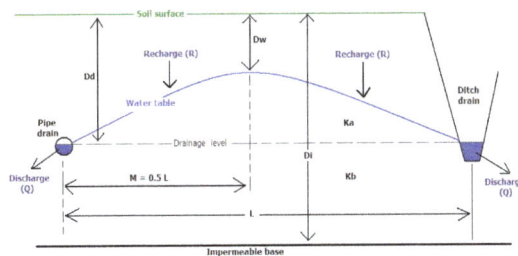

Geometry subsurface drainage system by pipes or ditches
D = depth   K = hydraulic conductivity   L = Drain spacing

Two-dimensional model of subsurface drainage in a vertical plane

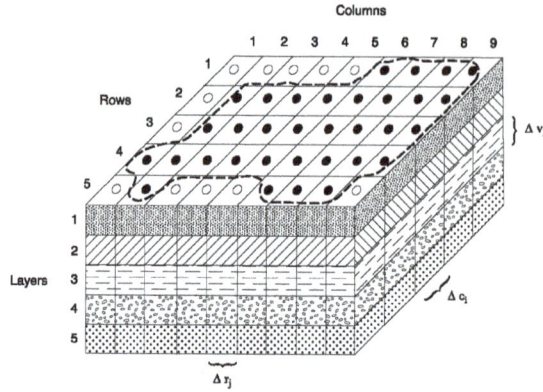

Three-dimensional grid, Modflow

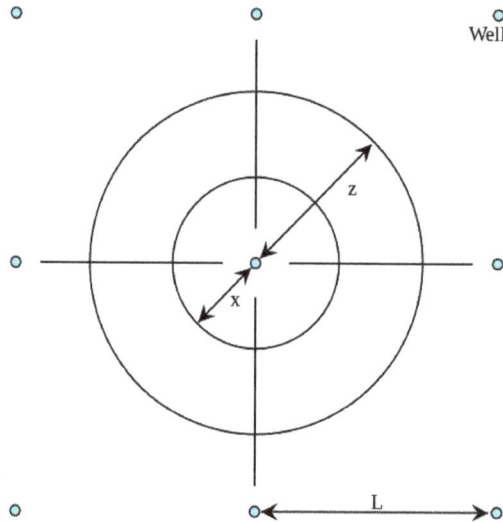

Map of a radial semi 3-dimensional model consisting of vertical
concentrical cylinders through which the flow passes radially to the well

Groundwater models can be one-dimensional, two-dimensional, three-dimensional and semi-three-dimensional. Two and three-dimensional models can take into account the anisotropy of the aquifer with respect to the hydraulic conductivity, i.e. this property may vary in different directions.

One-, Two- and Three-dimensional

1. One-dimensional models can be used for the vertical flow in a system of parallel horizontal layers.

2. Two-dimensional models apply to a vertical plane while it is assumed that the groundwater conditions repeat themselves in other parallel vertical planes (Fig. Two dimensional model of subsurface drainage in a vertical place). Spacing equations of subsurface drains and the groundwater energy balance applied to drainage equations are examples of two-dimensional groundwater models.

3. Three-dimensional models like Modflow require discretization of the entire flow domain. To that end the flow region must be subdivided into smaller elements (or cells), in both horizontal and vertical sense. Within each cell the parameters are maintained constant, but they may vary between the cells (Fig. Three dimensional grid, modflow). Using numerical solutions of groundwater flow equations, the flow of groundwater may be found as horizontal, vertical and, more often, as intermediate.

Semi Three-dimensional

In semi 3-dimensional models the horizontal flow is described by 2-dimensional flow equations (i. e. in horizontal x and y direction). Vertical flows (in z-direction) are described (a) with a 1-dimensional flow equation, or (b) derived from a water balance of horizontal flows converting the excess of horizontally incoming over the horizontally outgoing groundwater into vertical flow under the assumption that water is incompressible.

There are two classes of semi 3-dimensional models:

- *Continuous models* or *radial models* consisting of 2 dimensional submodels in vertical radial planes intersecting each other in one single axis. The flow pattern is repeated in each vertical plane fanning out from the central axis.

- *Discretized models* or *prismatic models* consisting of submodels formed by vertical blocks or prisms for the horizontal flow combined with one or more methods of superposition of the vertical flow.

Continuous Radial Model

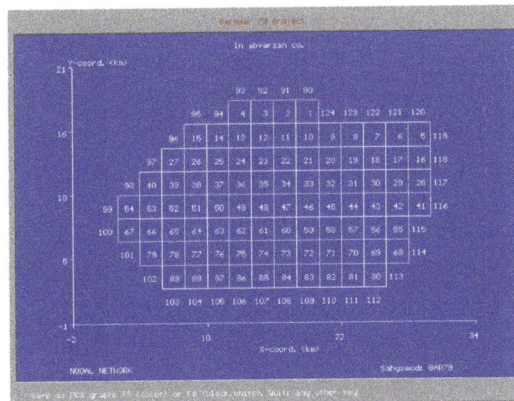

Map of a two-dimensional grid over an alluvial fan for a prismatic
semi 3-dimensional model, SahysMod

Een example of a non-discretized radial model is the description of groundwater flow moving radially towards a deep well in a network of wells from which water is abstracted. The radial flow passes through a vertical, cylindrical, cross-section representing the hydraulic equipotential of which the surface diminishes in the direction of the axis of intersection of the radial planes where the well is located (Fig. SahysMod).

Prismatically Discretized Model

Prismatically discretized models like SahysMod have a grid over the land surface only. The 2-dimensional grid network consists of triangles, squares, rectangles or polygons (Fig. SahysMod). Hence, the flow domain is subdivided into vertical blocks or prisms in figure under topic Boundary and Initial Conditions. The prisms can be discretized into *horizontal* layers with different characteristics that may also vary between the prisms. The groundwater flow between neighboring prisms is calculated using 2-dimensional horizontal groundwater flow equations. Vertical flows are found by applying one-dimensional flow equations in a vertical sense, or they can be derived from the water balance: excess of horizontal inflow over horizontal outflow (or vice versa) is translated into vertical flow.

In semi 3-dimensional models, intermediate flow between horizontal and vertical is not modelled like in truly 3-dimensional models. Yet, like the truly 3-dimensional models, such models do permit the introduction of horizontal and vertical subsurface drainage systems.

Semiconfined aquifers with a slowly permeable layer overlying the aquifer (the aquitard) can be included in the model by simulating vertical flow through it under influence of an overpressure in the aquifer proper relative to the level of the watertable inside or above the aquitard.

Groundwater Flow Equation

Used in hydrogeology, the groundwater flow equation is the mathematical relationship which is used to describe the flow of groundwater through an aquifer. The transient flow of groundwater is described by a form of the diffusion equation, similar to that used in heat transfer to describe the flow of heat in a solid (heat conduction). The steady-state flow of groundwater is described by a form of the Laplace equation, which is a form of potential flow and has analogs in numerous fields.

The groundwater flow equation is often derived for a small representative elemental volume (REV), where the properties of the medium are assumed to be effectively constant. A mass balance is done on the water flowing in and out of this small volume, the flux terms in the relationship being expressed in terms of head by using the constituitive equation called Darcy's law, which requires that the flow is slow.

Mass Balance

A mass balance must be performed, and used along with Darcy's law, to arrive at the transient groundwater flow equation. This balance is analogous to the energy balance used in heat transfer to arrive at the heat equation. It is simply a statement of accounting, that for a given control volume, aside from sources or sinks, mass cannot be created or destroyed. The conservation of mass states that, for a given increment of time (Δt), the difference between the mass flowing in across the boundaries, the mass flowing out across the boundaries, and the sources within the volume, is the change in storage.

$$\frac{\Delta M_{stor}}{\Delta t} = \frac{M_{in}}{\Delta t} - \frac{M_{out}}{\Delta t} - \frac{M_{gen}}{\Delta t}$$

Diffusion Equation (Transient Flow)

Mass can be represented as density times volume, and under most conditions, water can be considered incompressible (density does not depend on pressure). The mass fluxes across the boundaries then become volume fluxes (as are found in Darcy's law). Using Taylor series to represent the in and out flux terms across the boundaries of the control volume, and using the divergence theorem to turn the flux across the boundary into a flux over the entire volume, the final form of the groundwater flow equation (in differential form) is:

$$S_s \frac{\partial h}{\partial t} = -\nabla \cdot q - G.$$

This is known in other fields as the diffusion equation or heat equation, it is a parabolic partial differential equation (PDE). This mathematical statement indicates that the change in hydraulic head with time (left hand side) equals the negative divergence of the flux (q) and the source terms (G). This equation has both head and flux as unknowns, but Darcy's law relates flux to hydraulic heads, so substituting it in for the flux (q) leads to

$$S_s \frac{\partial h}{\partial t} = -\nabla \cdot (-K\nabla h) - G.$$

Now if hydraulic conductivity (K) is spatially uniform and isotropic (rather than a tensor), it can be taken out of the spatial derivative, simplifying them to the Laplacian, this makes the equation

$$S_s \frac{\partial h}{\partial t} = K\nabla^2 h - G.$$

Dividing through by the specific storage (S_s), puts hydraulic diffusivity ($\alpha = K/S_s$ or equivalently, $\alpha = T/S$) on the right hand side. The hydraulic diffusivity is proportional to the speed at which a finite pressure pulse will propagate through the system (large values of α lead to fast propagation of signals). The groundwater flow equation then becomes

$$\frac{\partial}{\partial} = \nabla h - G$$

Where the sink/source term, G, now has the same units but is divided by the appropriate storage term (as defined by the hydraulic diffusivity substitution).

Rectangular Cartesian Coordinates

Especially when using rectangular grid finite-difference models (*e.g.* MODFLOW, made by the USGS), we deal with Cartesian coordinates. In these coordinates the general Laplacian operator becomes (for three-dimensional flow) specifically

$$\frac{\partial h}{\partial t} = \alpha \left[\frac{\partial^2 h}{\partial x^2} + \frac{\partial^2 h}{\partial y^2} + \frac{\partial^2 h}{\partial z^2} \right] - G.$$

MODFLOW code discretizes and simulates an orthogonal 3-D form of the governing groundwater flow equation. However, it has an option to run in a "quasi-3D" mode if the user wishes to do so; in this case the model deals with the vertically averaged T and S, rather than k and S_s. In the quasi-3D mode, flow is calculated between 2D horizontal layers using the concept of leakage.

Circular Cylindrical Coordinates

Another useful coordinate system is 3D cylindrical coordinates (typically where a pumping well is a line source located at the origin — parallel to the z axis — causing converging radial flow). Under these conditions the above equation becomes (r being radial distance and θ being angle),

$$\frac{\partial h}{\partial t} = \alpha \left[\frac{\partial^2 h}{\partial r^2} + \frac{1}{r} \frac{\partial h}{\partial r} + \frac{1}{r^2} \frac{\partial^2 h}{\partial \theta^2} + \frac{\partial^2 h}{\partial z^2} \right] - G.$$

Assumptions

This equation represents flow to a pumping well (a sink of strength G), located at the origin. Both this equation and the Cartesian version above are the fundamental equation in groundwater flow, but to arrive at this point requires considerable simplification. Some of the main assumptions which went into both these equations are:

- the aquifer material is incompressible (no change in matrix due to changes in pressure — aka subsidence),

- the water is of constant density (incompressible),

- any external loads on the aquifer (e.g., overburden, atmospheric pressure) are constant,

- for the 1D radial problem the pumping well is fully penetrating a non-leaky aquifer,

- the groundwater is flowing slowly (Reynolds number less than unity), and

- the hydraulic conductivity (K) is an isotropic scalar.

Despite these large assumptions, the groundwater flow equation does a good job of representing the distribution of heads in aquifers due to a transient distribution of sources and sinks.

Laplace Equation (Steady-state Flow)

If the aquifer has recharging boundary conditions a steady-state may be reached (or it may be used as an approximation in many cases), and the diffusion equation (above) simplifies to the Laplace equation.

$$0 = \alpha \nabla^2 h$$

This equation states that hydraulic head is a harmonic function, and has many analogs in other fields. The Laplace equation can be solved using techniques, using similar assumptions stated above, but with the additional requirements of a steady-state flow field.

A common method for solution of this equations in civil engineering and soil mechanics is to use the graphical technique of drawing flownets; where contour lines of hydraulic head and the stream function make a curvilinear grid, allowing complex geometries to be solved approximately.

Steady-state flow to a pumping well (which never truly occurs, but is sometimes a useful approximation) is commonly called the Thiem solution.

Two-dimensional Groundwater Flow

The above groundwater flow equations are valid for three dimensional flow. In unconfined aquifers, the solution to the 3D form of the equation is complicated by the presence of a free surface water table boundary condition: in addition to solving for the spatial distribution of heads, the location of this surface is also an unknown. This is a non-linear problem, even though the governing equation is linear.

An alternative formulation of the groundwater flow equation may be obtained by invoking the Dupuit–Forchheimer assumption, where it is assumed that heads do not vary in the vertical direction (i.e., $\partial h / \partial z = 0$). A horizontal water balance is applied to a long vertical column with area $\delta x \delta y$ extending from the aquifer base to the unsaturated surface. This distance is referred to as the saturated thickness, b. In a confined aquifer, the saturated thickness is determined by the height of the aquifer, H, and the pressure head is non-zero everywhere. In an unconfined aquifer, the saturated thickness is defined as the vertical distance between the water table surface and the aquifer base. If $\partial h / \partial z = 0$, and the aquifer base is at the zero datum, then the unconfined saturated thickness is equal to the head, i.e., $b=h$.

Assuming both the hydraulic conductivity and the horizontal components of flow are uniform along the entire saturated thickness of the aquifer (i.e., $\partial q_x / \partial z = 0$ and $\partial K / \partial z = 0$), we can express Darcy's law in terms of integrated discharges, Q_x and Q_y:

$$Q_x = \int_0^b q_x dz = -Kb \frac{\partial h}{\partial x}$$

$$Q_y = \int_0^b q_y dz = -Kb \frac{\partial h}{\partial y}$$

Inserting these into our mass balance expression, we obtain the general 2D governing equation for incompressible saturated groundwater flow:

$$\frac{\partial nb}{\partial t} = \nabla \cdot (Kb\nabla h) + N.$$

Where n is the aquifer porosity. The source term, N (length per time), represents the addition of water in the vertical direction (e.g., recharge). By incorporating the correct definitions for saturated thickness, specific storage, and specific yield, we can transform this into two unique governing equations for confined and unconfined conditions:

$$S \frac{\partial h}{\partial t} = \nabla \cdot (KH\nabla h) + N.$$

(confined), where $S=S_s b$ is the aquifer storativity and

$$S_y \frac{\partial h}{\partial t} = \nabla \cdot (Kh\nabla h) + N.$$

(unconfined), where S_y is the specific yield of the aquifer.

Note that the partial differential equation in the unconfined case is non-linear, whereas it is linear in the confined case. For unconfined steady-state flow, this non-linearity may be removed by expressing the PDE in terms of the head squared:

$$\nabla \cdot (K\nabla h^2) = -2N.$$

Or, for homogeneous aquifers,

$$\nabla^2 h^2 = -\frac{2N}{K}.$$

This formulation allows us to apply standard methods for solving linear PDEs in the case of unconfined flow. For heterogeneous aquifers with no recharge, Potential flow methods may be applied for mixed confined/unconfined cases.

Water Table

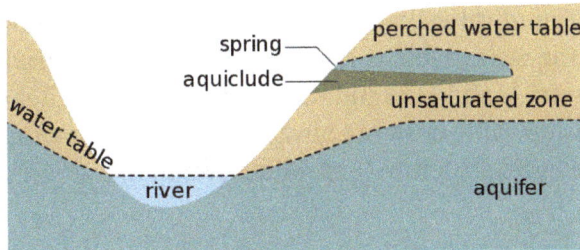

Cross section showing the water table varying with surface topography as well as a perched water table

The water table is the upper surface of the zone of saturation. The zone of saturation is where the pores and fractures of the ground are saturated with water.

The water table is the surface where the water pressure head is equal to the atmospheric pressure (where gauge pressure = 0). It may be visualized as the "surface" of the subsurface materials that are saturated with groundwater in a given vicinity.

The groundwater may be from precipitation or from groundwater flowing into the aquifer. In areas with sufficient precipitation, water infiltrates through pore spaces in the soil, passing through the unsaturated zone . At increasing depths water fills in more of the pore spaces in the soils, until a zone of saturation is reached. Below the water table, in the phreatic zone (zone of saturation), layers of permeable rock that yield groundwater are called aquifers. In less permeable soils, such as tight bedrock formations and historic lakebed deposits, the water table may be more difficult to define.

The water table should not be confused with the water level in a deeper well. If a deeper aquifer has a lower permeable unit that confines the upward flow, then the water level in this aquifer may rise to a level that is greater or less than the elevation of the actual water table. The elevation of the water in this deeper well is dependent upon the pressure in the deeper aquifer and is referred to as the potentiometric surface, not the water table.

Form

The water table may vary due to seasonal changes such as precipitation and evapotranspiration. In undeveloped regions with permeable soils that receive sufficient amounts of precipitation, the water table typically slopes toward rivers that act to drain the groundwater away and release the pressure in the aquifer. Springs, rivers, lakes and oases occur when the water table reaches the surface. Springs commonly form on hillsides, where the Earth's slanting surface may "intersect" with the water table. Groundwater entering rivers and lakes accounts for the base-flow water levels in water bodies.

Surface Topography

Within an aquifer, the water table is rarely horizontal, but reflects the surface relief due to the capillary effect (capillary fringe) in soils, sediments and other porous media. In the aquifer, groundwater flows from points of higher pressure to points of lower pressure, and the direction of groundwater flow typically has both a horizontal and a vertical component. The slope of the water table is known as the hydraulic gradient, which depends on the rate at which water is added to and removed from the aquifer and the permeability of the material. The water table does not always mimic the topography due to variations in the underlying geological structure (e.g., folded, faulted, fractured bedrock)

Perched Water Tables

A perched water table (or perched aquifer) is an aquifer that occurs above the regional water table, in the vadose zone. This occurs when there is an impermeable layer of rock or sediment (aquiclude) or relatively impermeable layer (aquitard) above the main water table/aquifer but below the surface of the land. If a perched aquifer's flow intersects the Earth's dry surface, at a valley wall for example, the water is discharged as a spring.

Fluctuations

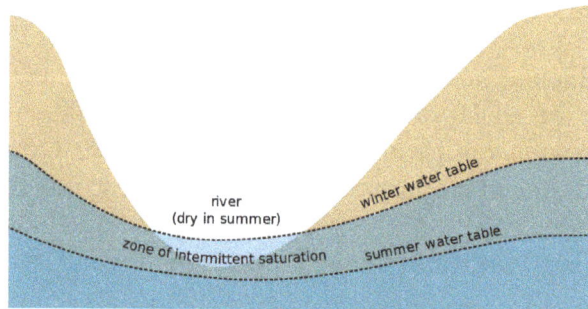

Seasonal fluctuations in the water table-during the dry season, river beds may dry up.

Tidal Fluctuations

On low-lying oceanic islands with porous soil, freshwater tends to collect in lenticular pools on top of the denser seawater intruding from the sides of the islands. Such an island's freshwater lens, and thus the water table, rises and falls with the tides.

Seasonal Fluctuations

In some regions, for example, Great Britain or California, winter precipitation is often higher than summer precipitation and so the groundwater storage is not fully recharged in summer. Consequently, the water table is lower during the summer. This disparity between the level of the winter and summer water table is known as the "zone of intermittent saturation", wherein the water table will fluctuate in response to climatic conditions.

Long-term Fluctuations

Fossil water is groundwater that has remained in an aquifer for several millennia and occurs mainly in deserts. It is non-renewable by present-day rainfall due to its depth below the surface, and any extraction causes a permanent change in the water table in such regions.

Effects on Climate

Aquifer drawdown or overdrafting and the pumping of fossil water may be a contributing factor to sea-level rise.

References

- Justin Gillis (April 26, 2012). "Study Indicates a Greater Threat of Extreme Weather". The New York Times. Retrieved April 27, 2012

- Rudolf Dvořák (2007). Extrasolar Planets. Wiley-VCH. pp. 139–140. ISBN 978-3-527-40671-5. Retrieved 2009-05-05

- "The Great Artesian Basin" (PDF). Facts: Water Series. Queensland Department of Natural Resources and Water. Archived from the original (PDF) on 2006-11-13. Retrieved 2007-01-03

- Post, V. E. A.; Groen, J.; Kooi, H.; Person, M.; Ge, S.; Edmunds, W. M. (2013). "Offshore fresh groundwater reserves as a global phenomenon". Nature. 504 (7478): 71–78. PMID 24305150. doi:10.1038/nature12858

- Metzger, Bruce M.; Coogan, Michael D. (1993). The Oxford Companion to the Bible. New York, NY: Oxford University Press. p. 369. ISBN 0195046455

- The World Bank, 2009 "Water and Climate Change: Understanding the Risks and Making Climate-Smart Investment Decisions". pp. xv. Retrieved 2011-10-24

- Evaristo, Jaivime; Jasechko, Scott; McDonnell, Jeffrey J. (2015-09-03). "Global separation of plant transpiration from groundwater and streamflow". Nature. NPG. 525 (7567): 91–94. Bibcode:2015Natur.525...91E. ISSN 0028-0836. PMID 26333467. doi:10.1038/nature14983

- Merrill, Eugene H.; Rooker, Mark F.; Grisanti, Michael A. (2011). The World and the Word. Nashville, TN: B&H Academic. p. 430. ISBN 9780805440317

- Paul J. Durack; Susan E. Wijffels & Richard J. Matear (27 April 2012). "Ocean Salinities Reveal Strong Global Water Cycle Intensification During 1950 to 2000". Science. 336 (6080): 455–458. Bibcode:2012Sci...336..455D. PMID 22539717. doi:10.1126/science.1212222

- U.S. Geologic Survey. GLACIER RETREAT IN GLACIER NATIONAL PARK, MONTANA. Retrieved on 2006-10-24

- Nick Strobel (June 12, 2010). "Planetary Science". Archived from the original on September 28, 2010. Retrieved September 28, 2010

Understanding Darcy's Law and Aquifer

The movement of groundwater can be understood in the context of Darcy's law. It is an equation that describes the flow of water through a porous media. This permeability is normally denoted in Darcy units. The law is mainly used to estimate water flow in aquifers. This chapter elucidates the crucial theories and principles of groundwater flow.

Darcy's Law

Darcy's law is an equation that describes the flow of a fluid through a porous medium. The law was formulated by Henry Darcy based on the results of experiments on the flow of water through beds of sand, forming the basis of hydrogeology, a branch of earth sciences.

Background

Although Darcy's law (an expression of Newton's second law) was determined experimentally by Darcy, it has since been derived from the Navier-Stokes equations via homogenization. It is analogous to Fourier's law in the field of heat conduction, Ohm's law in the field of electrical networks, or Fick's law in diffusion theory.

One application of Darcy's law is to analyze water flow through an aquifer; Darcy's law along with the equation of conservation of mass are equivalent to the groundwater flow equation, one of the basic relationships of hydrogeology.

Morris Muskat first refined Darcy's equation for single phase flow by including viscosity in the single (fluid) phase equation of Darcy, and this change made it suitable for the petroleum industry. Based on experimental results worked out by his colleagues Wyckoff and Botset, Muskat and Meres also generalized Darcy's law to cover multiphase flow of water, oil and gas in the porous medium of a petroleum reservoir. The generalized multiphase flow equations of Muskat et alios provide the analytical foundation for reservoir engineering that exists to this day.

Description

Darcy's law, as refined by Morris Muskat, at constant elevation is a simple proportional relationship between the instantaneous discharge rate through a porous medium, the viscosity of the fluid and the pressure drop over a given distance.

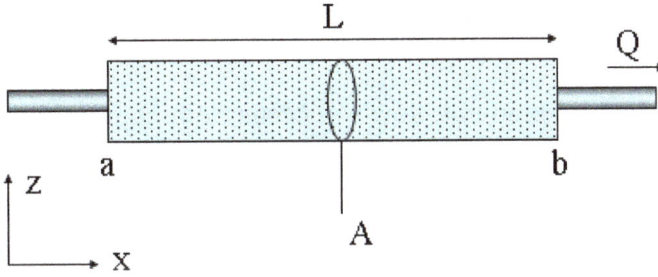

Diagram showing definitions and directions for Darcy's law.

$$Q = -\frac{\kappa A (p_b - p_a)}{\mu L}.$$

The above equation for single phase (fluid) flow is the defining equation for absolute permeability (single phase permeability). The total discharge, Q (units of volume per time, e.g., m³/s) is equal to the product of the intrinsic permeability of the medium, κ (m²), the cross-sectional area to flow, A (units of area, e.g., m²), and the total pressure drop $p_b - p_a$ (pascals), all divided by the viscosity, μ (Pa·s) and the length over which the pressure drop is taking place (L). The negative sign is needed because fluid flows from high pressure to low pressure. Note that the elevation head must be taken into account if the inlet and outlet are at different elevations. If the change in pressure is negative (where $p_a > p_b$), then the flow will be in the positive x direction. There have been several proposals for a constitutive equation for absolute permeability, and the most famous one is probably the Kozeny equation (also called Kozeny-Carman equation).

Dividing both sides of the equation by the area and using more general notation leads

$$q = -\frac{\kappa}{\mu} \nabla p,$$

where q is the flux (discharge per unit area, with units of length per time, m/s) and ∇p is the pressure gradient vector (Pa/m). This value of flux, often referred to as the Darcy flux or Darcy velocity, is not the velocity which the fluid traveling through the pores is experiencing. The fluid velocity (v) is related to the Darcy flux (q) by the porosity (φ). The flux is divided by porosity to account for the fact that only a fraction of the total formation volume is available for flow. The fluid velocity would be the velocity a conservative tracer would experience if carried by the fluid through the formation.

$$v = \frac{q}{\phi}.$$

Darcy's law is a simple mathematical statement which neatly summarizes several familiar properties that groundwater flowing in aquifers exhibits, including:

- if there is no pressure gradient over a distance, no flow occurs (these are hydrostatic conditions),

- if there is a pressure gradient, flow will occur from high pressure towards low pressure (opposite the direction of increasing gradient — hence the negative sign in Darcy's law),

- the greater the pressure gradient (through the same formation material), the greater the discharge rate, and the discharge rate of fluid will often be different — through different formation materials (or even through the same material, in a different direction) — even if the same pressure gradient exists in both cases.

A graphical illustration of the use of the steady-state groundwater flow equation (based on Darcy's law and the conservation of mass) is in the construction of flownets, to quantify the amount of groundwater flowing under a dam.

Darcy's law is only valid for slow, viscous flow; fortunately, most groundwater flow cases fall in this category. Typically any flow with a Reynolds number less than one is clearly laminar, and it would be valid to apply Darcy's law. Experimental tests have shown that flow regimes with Reynolds numbers up to 10 may still be Darcian, as in the case of groundwater flow. The Reynolds number (a dimensionless parameter) for porous media flow is typically expressed as

$$\text{Re} = \frac{\rho v d_{30}}{\mu},$$

where ρ is the density of water (units of mass per volume), v is the specific discharge (not the pore velocity — with units of length per time), d_{30} is a representative grain diameter for the porous media (often taken as the 30% passing size from a grain size analysis using sieves — with units of length), and μ is the viscosity of the fluid.

Derivation

For stationary, creeping, incompressible flow, i.e. $\dfrac{D(\rho u_i)}{Dt} \approx 0$, the Navier–Stokes equation simplifies to the Stokes equation:

$$\mu \nabla^2 u_i + \rho g_i - \partial_i p = 0,$$

where μ is the viscosity, u_i is the velocity in the i direction, g_i is the gravity component in the i direction and p is the pressure. Assuming the viscous resisting force is linear with the velocity we may write:

$$-\left(\kappa_{ij}\right)^{-1}\mu\phi u_j + \rho g_i - \partial_i p = 0,$$

where φ is the porosity, and κ_{ij} is the second order permeability tensor. This gives the velocity in the n direction,

$$\kappa_{ni}\left(\kappa_{ij}\right)^{-1}u_j = \delta_{nj}u_j = u_n = -\frac{\kappa_{ni}}{\phi\mu}\left(\partial_i p - \rho g_i\right),$$

which gives Darcy's law for the volumetric flux density in the n direction,

$$q_n = -\frac{\kappa_{ni}}{\mu}\left(\partial_i p - \rho g_i\right).$$

In isotropic porous media the off-diagonal elements in the permeability tensor are zero, $\kappa_{ij} = 0$ for $i \neq j$ and the diagonal elements are identical, $\kappa_{ii} = \kappa$, and the common form is obtained

$$\mathbf{q} = -\frac{\kappa}{\mu}\left(\nabla p - \rho\mathbf{g}\right).$$

The above equation is a governing equation for single phase fluid flow in a porous medium.

Additional Forms of Darcy's Law

Darcy's Law in Petroleum Engineering

Another derivation of Darcy's law is used extensively in petroleum engineering to determine the flow through permeable media — the most simple of which is for a one-dimensional, homogeneous rock formation with a single fluid phase and constant fluid viscosity.

$$Q = \frac{\kappa A}{\mu}\left(\frac{\partial p}{\partial x}\right),$$

where Q is the flowrate of the formation (in units of volume per unit time), k is the permeability of the formation (typically in millidarcys), A is the cross-sectional area of the formation, μ is the viscosity of the fluid (typically in units of centipoise). $\frac{\partial p}{\partial x}$ represents the pressure change per unit length of the formation. This equation can also be solved for permeability and is used to measure it, forcing a fluid of known viscosity through a core of a known length and area, and measuring the pressure drop across the length of the core.

Almost all oil reservoirs have a water zone blow the oil leg, and some have also a gas cap above the oil leg. When the reservoir pressure drops due to oil production, water flows into the oil zone from below, and gas flows into the oil zone from above (if the gas cap exists), and we get a simultaneous flow and immiscible mixing of all fluid phases in the oil zone. The operator of the oil field may also inject water (and/ or gas) in order to improve oil production. The petroleum industry is therefore using a generalized Darcy equation for multiphase flow that was developed by Muskat et alios. Because Darcy's name is so widespread and strongly associated with flow in porous media, the multiphase equation is denoted Darcy's law for multiphase flow or generalized Darcy equation (or law) or simply Darcy's equation (or law) or simply flow equation if the context says that the text is discussing the multiphase equation of Muskat et alios.

Darcy–Forchheimer Law

For flows in porous media with Reynolds numbers greater than about 1 to 10, inertial effects can also become significant. Sometimes an inertial term is added to the Darcy's equation, known as Forchheimer term. This term is able to account for the non-linear behavior of the pressure difference vs flow data.

$$\frac{\partial p}{\partial x} = -\frac{\mu}{\kappa}q - \frac{\rho}{\kappa_1}q^2,$$

where the additional term κ_1 is known as inertial permeability.

The flow in the middle of a sandstone reservoir is so slow that Forchheimer's equation is usually not needed, but the gas flow into a gas production well may be high enough to justify use of Forchheimer's equation. In this case the inflow performance calculations for the well, not the grid cell of the 3D model, is based on the Forchheimer equation. The effect of this is that an additional rate-dependent skin appears in the inflow performance formula.

Some carbonate reservoirs have lots of fractures, and Darcy's equation for multiphase flow is generalized in order to govern both flow in fractures and flow in the matrix (i.e. the traditional porous rock). The irregular surface of the fracture walls and high flow rate in the fractures, may justify use of Forchheimer's equation.

Darcy's Law for Gases in Fine Media (Knudsen Diffusion or Klinkenberg Effect)

For gas flow in small characteristic dimensions (e.g., very fine sand, nanoporous structures etc.), the particle-wall interactions become more frequent, giving rise to additional wall friction (Knudsen friction). For a flow in this region, where both viscous and Knudsen friction are present, a new formulation needs to be used. Knudsen presented a semi-empirical model for flow in transition regime based on his experiments on small capillaries. For a porous medium, the Knudsen equation can be given as:

$$N = -\left(\frac{\kappa}{\mu}\frac{p_a + p_b}{2} + D_K^{\text{eff}}\right)\frac{1}{R_g T}\frac{p_b - p_a}{L},$$

where N is the molar flux, R_g is the gas constant, T is the temperature, D_K^{eff} is the effective Knudsen diffusivity of the porous media. The model can also be derived from the first-principle-based binary friction model (BFM). The differential equation of transition flow in porous media based on BFM is given as:

$$\frac{\partial p}{\partial x} = -R_g T\left(\frac{\kappa p}{\mu} + D_K\right)^{-1} N.$$

This equation is valid for capillaries as well as porous media. The terminology of the Knudsen effect and Knudsen diffusivity is more common in mechanical and chemical engineering. In geological and petrochemical engineering, this effect is known as the Klinkenberg effect. Using the definition of molar flux, the above equation can be rewritten as:

$$\frac{\partial p}{\partial x} = -R_g T\left(\frac{\kappa p}{\mu} + D_K\right)^{-1}\frac{p}{R_g T}q.$$

This equation can be rearranged into the following equation:

$$q = -\frac{\kappa}{\mu}\left(1 + \frac{D_K \mu}{\kappa}\frac{1}{p}\right)\frac{\partial p}{\partial x}.$$

Comparing this equation with conventional Darcy's law, a new formulation can be given as:

$$q = -\frac{\kappa^{\text{eff}}}{\mu}\frac{\partial p}{\partial x},$$

where

$$\kappa^{\text{eff}} = \kappa\left(1 + \frac{D_K \mu}{\kappa}\frac{1}{p}\right).$$

This is equivalent to the effective permeability formulation proposed by Klinkenberg:

$$\kappa^{\text{eff}} = \kappa\left(1 + \frac{b}{p}\right).$$

where b is known as the Klinkenberg parameter, which depends on the gas and the porous medium structure. This is quite evident if we compare the above formulations. The Klinkenberg parameter b is dependent on permeability, Knudsen diffusivity and viscosity (i.e., both gas and porous medium properties).

Darcy's Law for Short Time Scales

For very short time scales, a time derivative of flux may be added to Darcy's law, which results in valid solutions at very small times (in heat transfer, this is called the modified form of Fourier's law),

$$\tau \frac{\partial q}{\partial t} + q = -\kappa \nabla h,$$

where τ is a very small time constant which causes this equation to reduce to the normal form of Darcy's law at "normal" times (> nanoseconds). The main reason for doing this is that the regular groundwater flow equation (diffusion equation) leads to singularities at constant head boundaries at very small times. This form is more mathematically rigorous, but leads to a hyperbolic groundwater flow equation, which is more difficult to solve and is only useful at very small times, typically out of the realm of practical use.

Brinkman form of Darcy's Law

Another extension to the traditional form of Darcy's law is the Brinkman term, which is used to account for transitional flow between boundaries (introduced by Brinkman in 1949),

$$\beta \nabla^2 q + q = -\frac{\kappa}{\mu} \nabla p,$$

where β is an effective viscosity term. This correction term accounts for flow through medium where the grains of the media are porous themselves, but is difficult to use, and is typically neglected.

Validity of Darcy's Law

Darcy's law is valid for laminar flow through sediments. In fine-grained sediments, the dimensions of interstices are small and thus flow is laminar. Coarse-grained sediments also behave similarly but in very coarse-grained sediments the flow may be turbulent. Hence Darcy's law is not always valid in such sediments. For flow through commercial pipes, the flow is laminar when Reynolds number is less than 2000, but in some sediments it has been found that flow is laminar when the value of Reynolds number is less than 1.

Darcy (Unit)

A darcy (or darcy unit) and millidarcy (md or mD) are units of permeability, named after Henry Darcy. They are not SI units, but they are widely used in petroleum engineering and geology. Like some other measures of permeability, a darcy has dimensional units in length^2.

Definition

Permeability measures the ability of fluids to flow through rock (or other porous media). The darcy is defined using Darcy's law, which can be written as:

$$Q = \frac{A\kappa\Delta P}{\mu\Delta x}$$

where:

Q is the volumetric fluid flow rate through the medium

A is the area of the medium

κ is the permeability of the medium

μ is the dynamic viscosity of the fluid

ΔP is the applied pressure difference

Δx is the thickness of the medium

The darcy is referenced to a mixture of unit systems. A medium with a permeability of 1 darcy permits a flow of 1 cm^3/s of a fluid with viscosity 1 cP (1 mPa·s) under a pressure gradient of 1 atm/cm acting across an area of 1 cm^2.

Typical values of permeability range as high as 100,000 darcys for gravel, to less than 0.01 microdarcy for granite. Sand has a permeability of approximately 1 darcy.

Origin

The darcy is named after Henry Darcy. Rock permeability is usually expressed in millidarcys (md) because rocks hosting hydrocarbon or water accumulations typically exhibit permeability ranging from 5 to 500 md.

The odd combination of units comes from Darcy's original studies of water flow through columns of sand. Water has a viscosity of 1.0019 cP at about room temperature.

The unit is named after Henry Darcy, and the unit abbreviation is not capitalized (contrary to industry use). The American Association of Petroleum Geologists use the following unit abbreviations and grammar in their publications:

- darcy (plural darcys, not darcies): d

- millidarcy (plural millidarcys, not millidarcies): md

Conversions

Converted to SI units, 1 darcy is equivalent to 9.869233×10^{-13} m² or 0.9869233 (μm)². This conversion is usually approximated as 1 (μm)². Note that this is the reciprocal of 1.013250—the conversion factor from atmospheres to bars.

Specifically in the hydrology domain, permeability of soil or rock may also be defined as the flux of water under hydrostatic pressure ($\sim$ 0.1 bar/m) at a temperature of 20°C. In this specific setup, 1 darcy is equivalent to 0.831 m/day.

Hydraulic Conductivity

In case of horizontally stratified soil, the hydraulic conductivity of different layer may be different. However, if we consider a particular layer, the layer is homogeneous and isotropic in nature. For this type of field problem, it is always possible to obtain an equivalent hydraulic conductivity for the whole aquifer which will produce the same discharge as that of the stratified aquifer with different hydraulic conductivities.

Consider an aquifer consisting with n horizontal layers as shown in figure below. Each layer is individually isotropic in nature. Let the thickness of the layers are dz_1, dz_2, dz_3.............dz_n and hydraulic conductivity of the layers are K_1, K_2, K_3.............K_n

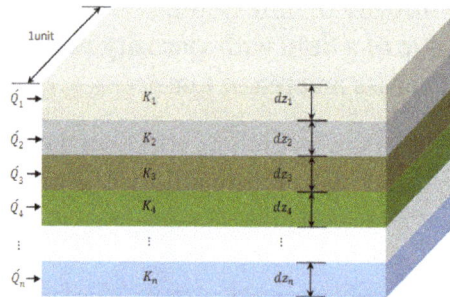

Aquifer with horizontal strata of different hydraulic conductivities

The discharge per unit width through the first layer may be written as

$$\dot{Q}_1 = -\left(dz_1 \times 1\right) K_1 \frac{\partial \varphi}{\partial x}$$

$$= -K_1 dz_1 \frac{\partial \varphi}{\partial x}$$

Similarly

$$\dot{Q}_2 = -K_2 dz_2 \frac{\partial \varphi}{\partial x}$$

$$\dot{Q}_3 = -K_3 dz_3 \frac{\partial \varphi}{\partial x}$$

And

$$\dot{Q}_3 = -K_3 dz_3 \frac{\partial \varphi}{\partial x}$$

The total horizontal flow through the aquifer is

$$\dot{Q}_x = \dot{Q}_1 + \dot{Q}_2 + \dot{Q}_3 + \ldots \ldots \ldots \dot{Q}_n$$

$$= -\left[K_1 dz_1 \frac{\partial \varphi}{\partial \varphi} + K_2 dz_2 \frac{\partial \varphi}{\partial \varphi} + K_3 dz_3 \frac{\partial \varphi}{\partial \varphi} + \ldots \ldots \ldots \ldots \ldots \ldots + K_n dz_n \frac{\partial \varphi}{\partial \varphi} \right]$$

$$= -\frac{d\varphi}{dx} [K_1 dz_1 + K_2 dz_2 + K_3 dz_3 + \ldots \ldots \ldots + K_n dz_n]$$

If we consider K_x as the equivalent hydraulic conductivity of the aquifer the total flow through the aquifer is

$$\dot{Q}_x = -K_x \left(dz_1 + dz_2 + dz_3 + \ldots \ldots \ldots dz_n \right) \frac{\partial \varphi}{\partial x}$$

Equating the above two:

$$-K_x \left(dz_1 + dz_2 + dz_3 + \ldots + dz_n \right) \frac{\partial \varphi}{\partial x} = -\frac{\partial \varphi}{\partial x} [K_1 dz_1 + K_2 dz_2 + \ldots K_n dz_n]$$

$$\Rightarrow K_x = \frac{K_1 dz_1 + K_2 dz_2 + \ldots + K_n dz_n}{dz_1 + dz_2 + \ldots dz_n}$$

$$\Rightarrow K_x = \frac{\sum_{i=1}^{n} K_i dz_i}{\sum_{i=1}^{n} dz_i}$$

$$\Rightarrow K_x = \frac{\sum_{i=1}^{n} K_i dz_i}{H}$$

Where, $H = dz_1, dz_2, dz_3 \ldots \ldots \ldots dz_n$, total depth of the aquifer.

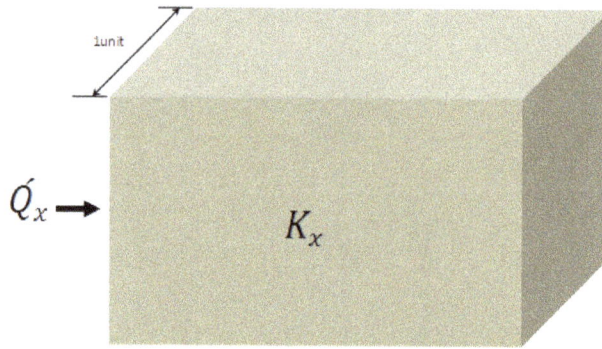

Equivalent homogeneous aquifer

The aquifer medium can now be replaced by a homogeneous medium with horizontal hydraulic conductivity of K_x as shown in figure above. In the above calculation of equivalent horizontal hydraulic conductivity, the strata are parallel to the flow direction. In some cases, the strata may be perpendicular to the flow direction. Figure below shows an aquifer, where strata are perpendicular to the flow direction.

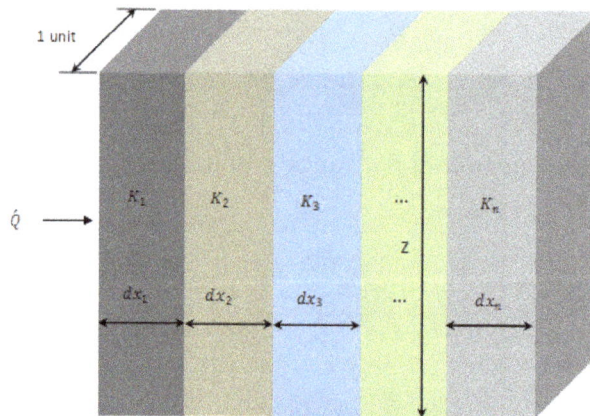

Aquifer with vertical strata of different hydraulic conductivities

In this case, the flow per unit width of the aquifer can be written as

$$\acute{Q} = -K_1(z \times 1)\frac{d\varphi_1}{dx_1}$$

$$= -K_1 z \frac{d\varphi_1}{dx_1} \Rightarrow d\varphi_1 = -\frac{dx_1}{K_1 z}\acute{Q}$$

Where $d\varphi_1$ is the head loss within the first layer.

For continuity, the flow $\acute{Q}$ must be same for the other layers also. As such, the head in other layers are

$$d\varphi_2 = -\frac{dx_2}{K_2 z}\dot{Q}$$

$$d\varphi_3 = -\frac{dx_3}{K_3 z}\dot{Q}$$

And

$$d\varphi_n = -\frac{dx_n}{K_n z}\dot{Q}$$

The total head loss through the aquifer is

$$d\varphi = d\varphi_1 + d\varphi_2 + d\varphi_3 + \ldots + d\varphi_n$$

$$= -\frac{\dot{Q}}{z}\left[\frac{dx_1}{K_1} + \frac{dx_2}{K_2} + \frac{dx_3}{K_3} + \ldots + \frac{dx_n}{K_n}\right]$$

If we consider the aquifer as homogeneous with equivalent hydraulic conductivity at K_x, the total head loss would be,

$$d\varphi = -\frac{\dot{Q}}{z}\left[\frac{dx_1 + dx_2 + \ldots\ldots\ldots + dx_n}{K_x}\right]$$

Equating the above two

$$-\frac{\dot{Q}}{z}\left[\frac{dx_1 + dx_2 + \ldots\ldots\ldots + dx_n}{K_x}\right] = -\frac{\dot{Q}}{z}\left[\frac{dx_1}{K_1} + \frac{dx_2}{K_2} + \ldots\ldots\ldots + \frac{dx_n}{K_n}\right]$$

Or

$$K_x = \frac{dx_1 + dx_2 + \ldots + dx_n}{\frac{dx_1}{K_1} + \frac{dx_2}{K_2} + \ldots + \frac{dx_n}{K_n}}$$

Or

$$K_x = \frac{\sum_{i=1}^{n} dx_i}{\sum_{i=1}^{n} \frac{dx_i}{K_i}}$$

Or

$$K_x = \frac{L}{\sum_{i=1}^{n} \frac{dx_i}{K_i}}$$

Where, L is the length of the aquifer.

The aquifer medium can now be replaced by a homogeneous medium with horizontal hydraulic conductivity of K_x as shown below

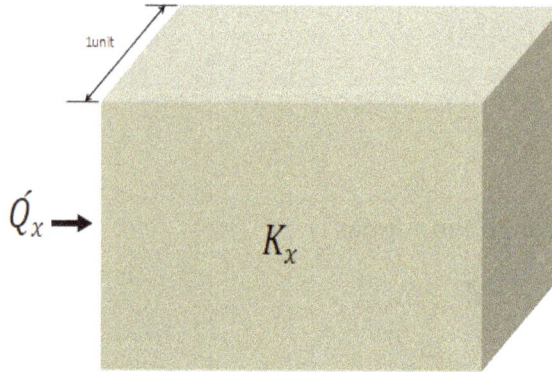

Equivalent homogeneous aquifer

Aquifer

Consider the flow through a confined aquifer as shown in the figure below. The width of the aquifer is W. The depth of the aquifer is B. The total discharge in the x direction through the area WB can be written as,

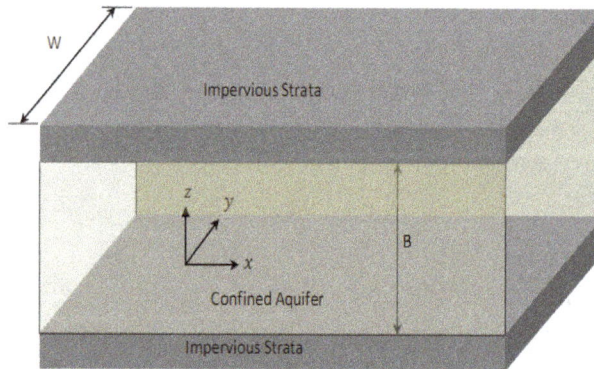

A confined Aquifer

The discharge per unit width through the first Layer may be written as

$$Q_x = WB\left(-K_{xx}\frac{\partial\varphi}{\partial x} - K_{xy}\frac{\partial\varphi}{\partial y}\right)$$

$$= -WBK_{xx}\frac{\partial\varphi}{\partial x} - WBK_{xy}\frac{\partial\varphi}{\partial y}$$

The discharge per unit width of the aquifer can be written as,

$$Q_x = -\left(BK_{xx}\right)\frac{\partial\varphi}{\partial x} - \left(BK_{xy}\right)\frac{\partial\varphi}{\partial y}$$

Putting $T_{xx} = BK_{xx}$ and $T_{xy} = BK_{xy}{}'$, the above equation becomes

$$\dot{Q}_x = -T_{xx}\frac{\partial\varphi}{\partial x} - T_{xy}\frac{\partial\varphi}{\partial y}$$

Similarly in the direction of y the discharge per unit width of the aquifer can be written as

$$\dot{Q}_y = -\left(BK_{yy}\right)\frac{\partial\varphi}{\partial y} - \left(BK_{yx}\right)\frac{\partial\varphi}{\partial x}$$

Putting $T_{yy} = BK_{yy}$ and $T_{yx} = BK_{yx}{}'$, the abvoe equation becomes

$$\dot{Q}_y = -T_{yy}\frac{\partial\varphi}{\partial y} - T_{yx}\frac{\partial\varphi}{\partial x}$$

In matrix form, it may be written as,

$$\begin{Bmatrix} \dot{Q}_x \\ \dot{Q}_y \end{Bmatrix} = \begin{bmatrix} T_{xx} & T_{xy} \\ T_{yx} & T_{yy} \end{bmatrix} \begin{bmatrix} -\dfrac{\partial\varphi}{\partial x} \\ -\dfrac{\partial\varphi}{\partial x} \end{bmatrix}$$

And in vector form, we can write as

$$\dot{Q} = -T\nabla\phi$$

Where T is the transmissivity of the aquifer which represents the discharge through the entire thickness of the aquifer under unit hydraulic gradient.

In case of homogeneous isotropic porous medium, the discharge per unit width of the aquifer may be written as,

$$\left. \begin{aligned} Q_x &= -T\frac{\partial\varphi}{\partial x} \\ Q_y &= -T\frac{\partial\varphi}{\partial y} \end{aligned} \right\}$$

When principal directions are used as the coordinate system, the equation above can be written as,

$$\begin{Bmatrix} \dot{Q}_x \\ \dot{Q}_y \end{Bmatrix} = \begin{bmatrix} T'_{\dot{x}\dot{x}} & 0 \\ 0 & T'_{\dot{y}\dot{y}} \end{bmatrix} \begin{Bmatrix} -\dfrac{\partial \varphi}{\partial x} \\ -\dfrac{\partial \varphi}{\partial y} \end{Bmatrix}$$

Or

$$\left.\begin{aligned} \dot{Q}_x &= -T'_{\dot{x}\dot{x}} \dfrac{\partial \varphi}{\partial x} \\ \dot{Q}_y &= -T'_{\dot{y}\dot{y}} \dfrac{\partial \varphi}{\partial y} \end{aligned}\right\}$$

Where $\dot{x}\,\dot{y}$ is the principal coordinate.

The transformation from any arbitrary xy coordinate system to the principal $\dot{x}\,\dot{y}$ coordinate system can be obtained using the following relationship

$$T_{\dot{x}\dot{x}} = \frac{T_{xx} + T_{yy}}{2} + \left[\left(\frac{T_{xx} - T_{yy}}{2} \right)^2 + T_{xy}^2 \right]^{1/2}$$

$$T_{\dot{y}\dot{y}} = \frac{T_{xx} + T_{yy}}{2} + \left[\left(\frac{T_{xx} - T_{yy}}{2} \right)^2 + T_{xy}^2 \right]^{1/2}$$

Specific Storage

In the field of hydrogeology, "storage properties" are physical properties that characterize the capacity of an aquifer to release groundwater. These properties are Storativity (S), specific storage (S_s) and specific yield (S_y).

They are often determined using some combination of field tests (e.g., aquifer tests) and laboratory tests on aquifer material samples.

Storativity

Storativity or the storage coefficient is the volume of water released from storage per unit decline in hydraulic head in the aquifer, per unit area of the aquifer. Storativity is a dimensionless quantity, and ranges between 0 and the effective porosity of the aquifer.

$$S = \frac{dV_w}{dh} \frac{1}{A} = S_s b + S_y$$

- V_w is the volume of water released from storage ($[L^3]$);

- h is the hydraulic head ([L])

- S_s is the specific storage

- S_y is the specific yield

- b is the thickness of aquifer

- A is the area ([L²])

Confined

For a confined aquifer or aquitard, storativity is the vertically integrated specific storage value. Therefore, if the aquitard is homogeneous:

$$S = S_s b$$

Unconfined

For unconfined aquifer storativity is approximately equal to the specific yield (S_y) since the release from specific storage (S_s) is typically orders of magnitude less ($S_s b \ll S_y$).

$$S = S_y$$

The specific storage is the amount of water that a portion of an aquifer releases from storage, per unit mass or volume of aquifer, per unit change in hydraulic head, while remaining fully saturated.

Mass specific storage is the mass of water that an aquifer releases from storage, per mass of aquifer, per unit decline in hydraulic head:

$$(S_s)_m = \frac{1}{m_a} \frac{dm_w}{dh}$$

where

$(S_s)_m$ is the mass specific storage ([L^{-1}]);

m_a is the mass of that portion of the aquifer from which the water is released ([M]);

dm_w is the mass of water released from storage ([M]); and

dh is the decline in hydraulic head ([L]).

Volumetric specific storage (or volume specific storage) is the volume of water that an aquifer releases from storage, per volume of aquifer, per unit decline in hydraulic head (Freeze and Cherry, 1979):

$$S_s = \frac{1}{V_a}\frac{dV_w}{dh} = \frac{1}{V_a}\frac{dV_w}{dp}\frac{dp}{dh} = \frac{1}{V_a}\frac{dV_w}{dp}\gamma_w$$

where

> S_s is the volumetric specific storage ($[L^{-1}]$);
>
> V_a is the bulk volume of that portion of the aquifer from which the water is released ($[L^3]$);
>
> dV_w is the volume of water released from storage ($[L^3]$);
>
> dp is the decline in pressure($N{\bullet}m^{-2}$ or $[ML^{-1}T^{-2}]$) ;
>
> dh is the decline in hydraulic head ($[L]$) and
>
> γ_w is the specific weight of water ($N{\bullet}m^{-3}$ or $[ML^{-2}T^{-2}]$).

In hydrogeology, volumetric specific storage is much more commonly encountered than mass specific storage. Consequently, the term specific storage generally refers to volumetric specific storage.

In terms of measurable physical properties, specific storage can be expressed as

$$S_s = \gamma_w(\beta_p + n \cdot \beta_w)$$

where

> γ_w is the specific weight of water ($N{\bullet}m^{-3}$ or $[ML^{-2}T^{-2}]$)
>
> is the porosity of the material (dimensionless ratio between 0 and 1)
>
> β_p is the compressibility of the bulk aquifer material (m^2N^{-1} or $[LM^{-1}T^2]$), and
>
> β_w is the compressibility of water (m^2N^{-1} or $[LM^{-1}T^2]$)

The compressibility terms relate a given change in stress to a change in volume (a strain). These two terms can be defined as:

$$\beta_p = -\frac{dV_t}{d\sigma_e}\frac{1}{V_t} \qquad\qquad \beta_w = -\frac{dV_w}{dp}\frac{1}{V_w}$$

where

> σ_e is the effective stress (N/m^2 or $[MLT^{-2}/L^2]$)

These equations relate a change in total or water volume (σ_e or V_w) per change in applied stress (effective stress $-\sigma_e$ or pore pressure $- p$) per unit volume. The compressibilities (and therefore also S_s) can be estimated from laboratory consolidation tests (in an apparatus called a consolidometer), using the consolidation theory of soil mechanics (developed by Karl Terzaghi).

Specific Yield

Specific yield, also known as the drainable porosity, is a ratio, less than or equal to the effective porosity, indicating the volumetric fraction of the bulk aquifer volume that a given aquifer will yield when all the water is allowed to drain out of it under the forces of gravity:

$$S_y = \frac{V_{wd}}{V_T}$$

where

V_{wd} is the volume of water drained, and

V_T is the total rock or material volume

It is primarily used for unconfined aquifers, since the elastic storage component, S_s, is relatively small and usually has an insignificant contribution. Specific yield can be close to effective porosity, but there are several subtle things which make this value more complicated than it seems. Some water always remains in the formation, even after drainage; it clings to the grains of sand and clay in the formation. Also, the value of specific yield may not be fully realized for a very long time, due to complications caused by unsaturated flow.

Phreatic Aquifer

A phreatic aquifer is defined as the aquifer where water table is considered as the upper boundary of the aquifer. In case of phreatic aquifer, the phreatic surface is never horizontal and the equipotential lines are not vertical. As such, the hydraulic head (φ) is a function of spatial coordinates x, y, z and time t. Further on the phreatic surface, a non-linear boundary condition has to be specified. At the same time, the location of the phreatic surface is also not known. Figure (a) shows the actual phreatic surface with streamline. At point p, specific discharge is in a direction tangent to the streamline. The specific discharge can be expressed as,

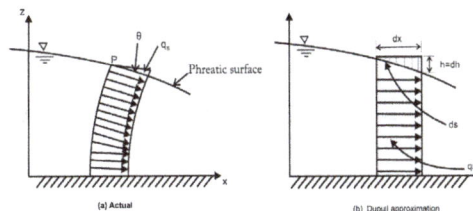

(a) Actual   (b) Dupui approximation

Pressure distribution

$$q_s = -K\frac{d\varphi}{ds} = -K\frac{dz}{ds} = -K\sin\theta$$

At phreatic surface, pressure is zero, hence $\varphi = z$. As per the observation made by Dupuit (1863), the slope of the phreatic surface is very small and is in the range of 1 in 1000 to 10 in 1000. As such, Dupuit suggested that sinθ can be replaced by tanθ. Thus the above equation can be written as

$$\left.\begin{aligned} q_s &= -K\tan\theta \\ &= -K\frac{dh}{dx} \end{aligned}\right\}$$

It may be noted that assumption of small θ is equivalent to the following assumptions

1. Equipotential lines are vertical, and

2. The flow is essentially horizontal

For the figure below, if we consider Dupuit assumption, the hydraulic head is a function of x and y only, *i.e.* h(x,y).

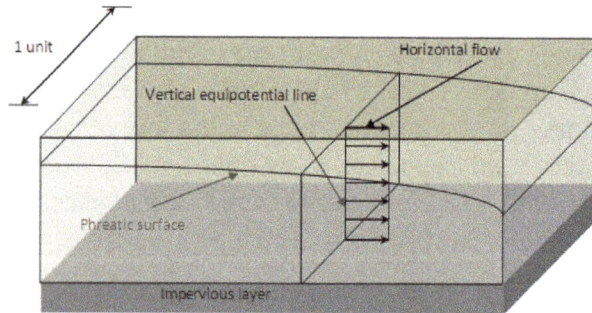

Flow in phreatic aquifer

Thus

$$q_x = -K\frac{\partial h}{\partial x}$$

And

$$q_y = -K\frac{\partial h}{\partial y}$$

The total discharge per unit width of the aquifer can be written as

$$\dot{Q}_x = -K(1\times h)\frac{\partial h}{\partial x}$$

$$= -Kh\frac{\partial h}{\partial x} = -K\frac{\partial}{\partial x}\left(\frac{h^2}{2}\right)$$

And

$$\dot{Q}_y = -Kh\frac{\partial h}{\partial y} = -K\frac{\partial}{\partial y}\left(\frac{h^2}{2}\right)$$

Discharge Between Two Reservoirs

Considering Dupuit assumption of horizontal flow, the discharge between two reservoirs can be calculated easily. Consider the phreatic aquifer shown in figure below. The hydraulic conductivity of the homogeneous aquifer is K. The discharge per unit width of the aquifer in x direction can be obtained by

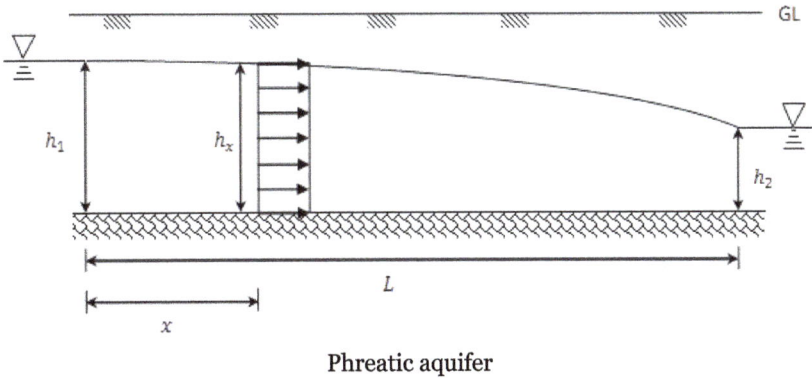

Phreatic aquifer

$$\dot{Q}_x = -Kh(x)\frac{\partial h}{\partial x}$$

$$\Rightarrow \dot{Q}_x' dx = -Khdh$$

Integrating between x = 0 to any distance x

$$At\ x = 0 \qquad\qquad h = h_1$$

$$At\ x = L \qquad\qquad h = h_2$$

$$\dot{Q}\int_{x=0}^{x=L}\partial x = -k\int_{h_1}^{h_2}h\partial h$$

$$\Rightarrow \dot{Q}x\Big]_0^L = --K\frac{h^2}{2}\Big]_{h_1}^{h_2}$$

$$\Rightarrow \dot{Q}L = K\frac{h_1^2 - h_2^2}{2}$$

$$Or\ \dot{Q} = K\frac{h_1^2 - h_2^2}{2L}$$

Equation above can be used to calculate discharge through the aquifer. This equation is also known as the Dupuit-Forchheimer discharge formula.

Horizontal Stratified Aquifer

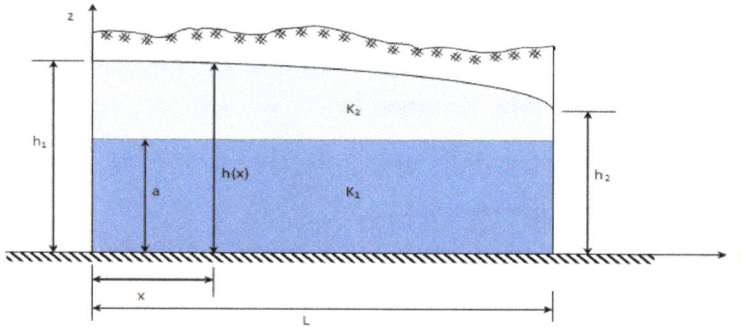

Phreatic aquifer with two horizontal strata

Consider the horizontally stratified aquifer shown in figure above. The hydraulic conductivity of the upper strata of the unconfined aquifer is K_2 and that for the bottom strata is K_1. The depth of the bottom strata is a. Considering Dupuit assumption, the discharge per unit width at a distance x from the left reservoir can be expressed as,

$$\dot{Q} = -K_1 a\frac{dh}{dx} - K_2 (h(x) - a)\left(\frac{dh}{dx}\right)$$

Integrating from $x = 0$, $h(x) = h_1$ and $x = L$, $h(x) = h_2$, we can obtain

$$\int_0^L \dot{Q}\ dx = -K_1 a\int_{h_1}^{h_2} dh - K_2\int_{h_1}^{h_2} (h(x) - a)dh$$

$$\Rightarrow \dot{Q}L = K_1 a(h_1 - h_2) + K_2\frac{h_1^2 - h_2^2}{2} = K_2 a(h_1 - h_2)$$

$$\Rightarrow \dot{Q}L = \frac{k_2}{2}(h_1 - h_2)\left[\frac{2aK_1}{K_2} + h_1 + h_2 - 2a\right]$$

$$\Rightarrow \dot{Q} = \frac{k_2}{2L}(h_1 - h_2)\left[h_1 + h_2 - 2a + 2a\left(\frac{K_1}{K_2}\right)\right]$$

Equation above can be used to calculate the discharge through the horizontal stratified unconfined aquifer.

In case of n numbers of strata, the equation can be expressed as

$$\dot{Q} = \frac{K_n}{2L}(h_1 - h_2)\left[(h_1 + h_2) - 2(a_1 + a_2 + \ldots + a_{n-1}) + 2\left\{\left(\frac{K_2}{K_n}\right)a_1 + \ldots + \left(\frac{K_{n-1}}{K_n}\right)a_{n-1}\right\}\right]$$

Flow Through Unconfined Vertically Stratified Aquifer

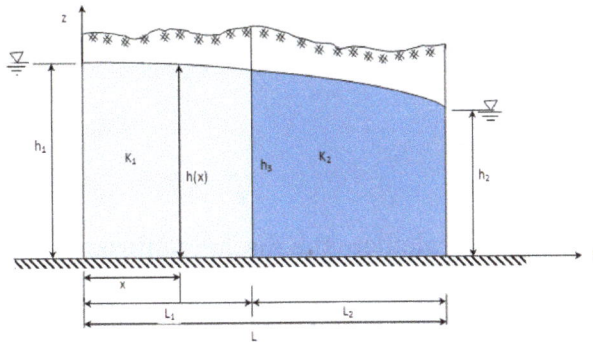

Phreatic aquifer with two vertical strata

Consider the vertically stratified aquifer as shown in figure above. The left strata has the hydraulic conductivity of K_1 and the right strata has the hydraulic conductivity of K_2. If we consider the first strata

$$\dot{Q} = -K_1 h(x)\frac{dh}{dx}$$

Integrating from $x = 0$, $h(x) = h_1$ and $x = L_1$, $h(x) = h_3$, we can obtain,

$$\dot{Q} = -\frac{K_1}{2L_1}\left(h_1^2 - h_3^2\right)$$

$$h_3^2 = h_2^2 - \frac{2\dot{Q}L_2}{K_2}$$

Similarly for the second strata, we can obtain

$$h_3^2 = h_2^2 - \frac{2\dot{Q}L_2}{K_2}$$

Equating the above two equations:

$$h_1^2 - \frac{2\dot{Q}L_1}{K_1} = h_2^2 - \frac{2\dot{Q}L_2}{K_2}$$

$$h_1^2 - h_2^2 = 2\dot{Q}\left(\frac{L_1}{K_1} + \frac{L_2}{K_2}\right)$$

$$\dot{Q} = \frac{h_1^2 - h_2^2}{2\left(\dfrac{L_1}{K_1} + \dfrac{L_2}{K_2}\right)}$$

The equation can be used for calculating discharge through the vertically stratified unconfined aquifer.

In case of n numbers of strata, the equation can be expressed as

$$\dot{Q} = \frac{h_1^2 - h_2^2}{2\sum_{i=1}^{n}\dfrac{L_i}{K_i}}$$

Equation above can be used to calculate the discharge through the vertically stratified unconfined aquifer.

Two-dimensional Flow in Confined Aquifer'

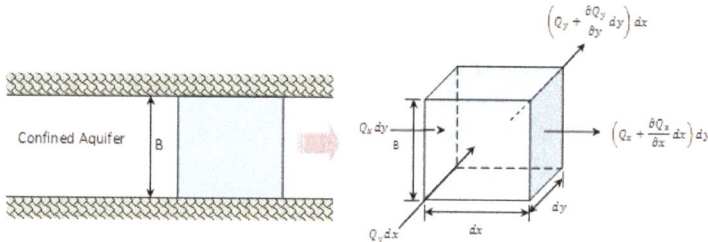

Confined aquifer with elementary control volume

Let, B be the thickness of the aquifer.

Q_x is the inflow per unit width in x direction

Q_y is the inflow per unit width in y direction

Inflow in x direction is

$Q_x dy$

Outflow in x direction is

$$\left(Q_x + \frac{\partial Q_x}{\partial x}dx\right)dy$$

Net flow in x direction

$$= Q_x dy - \left(Q_x + \frac{\partial Q_x}{\partial x} dx \right) dy$$

$$= -\frac{\partial Q_x}{\partial x} dx dy$$

Similarly, net flow in y direction is

$$-\frac{\partial Q_y}{\partial y} dx dy$$

Total net flow through the control volume is

$$\left(-\frac{\partial Q_x}{\partial x} - \frac{\partial Q_y}{\partial y} \right) dx dy$$

Now, as we know that aquifer storativity

$$S_S = \frac{\Delta V_\omega}{A \Delta \varphi}$$

$$\Rightarrow \Delta V_\omega = S_S A \Delta \varphi$$

$$= S_S dx dy \Delta \varphi$$

Change in storage in time dt is

$$S_S \frac{\partial \varphi}{\partial t} dx dy$$

Total net flow = change in storage in time dt

$$\left(-\frac{\partial Q_x}{\partial x} - \frac{\partial Q_y}{\partial y} \right) dx dy = S_S \frac{\partial \varphi}{\partial t} dx dy$$

$$\Rightarrow -\left(\frac{\partial Q_x}{\partial x} + \frac{\partial Q_y}{\partial y} \right) = S_S \frac{\partial \varphi}{\partial t}$$

Now putting,

$$Q_x = -T_x \frac{\partial \varphi}{\partial x} \text{ and } Q_y = -T_y \frac{\partial \varphi}{\partial y}$$

We have,

$$\frac{\partial}{\partial x}\left(T_x\frac{\partial\varphi}{\partial x}\right) + \frac{\partial}{\partial y}\left(T_y\frac{\partial\varphi}{\partial y}\right) = S_s\frac{\partial\varphi}{\partial t}$$

This is the flow equation for anisotropic non-homogeneous confined aquifer for unsteady condition.

In case of homogeneous aquifer, the equation becomes

$$T_x\frac{\partial^2\varphi}{\partial x^2} + T_y\frac{\partial^2\varphi}{\partial y^2} = S_s\frac{\partial\varphi}{\partial t}$$

In case of homogeneous isotropic aquifer $T_x = T_y = T$

$$\frac{\partial^2\varphi}{\partial x^2} + \frac{\partial^2\varphi}{\partial y^2} = \frac{S_s}{T}\frac{\partial\varphi}{\partial t}$$

If a source N(x, y, t) is present, the equation becomes

$$\frac{\partial}{\partial x}\left(T_x\frac{\partial\varphi}{\partial x}\right) + \frac{\partial}{\partial y}\left(T_y\frac{\partial\varphi}{\partial y}\right) + N\left(x,y,t\right) = S_s\frac{\partial\varphi}{\partial t}$$

Here, the source term represents the pumping or recharge per unit horizontal area of the aquifer. If case of pumping a negative sign is used as mass is withdrawn from the control volume. On the other hand, the source term will be positive in case of recharge as we are adding mass to the system. For example, if Qm^3/sec is the pumping rate from the control volume, the value of N(x, y, t) will be $-\dfrac{Q}{\left(\Delta x\Delta y\right)}m/\sec.$

Two-dimensional Flow in a Phreatic Aquifer

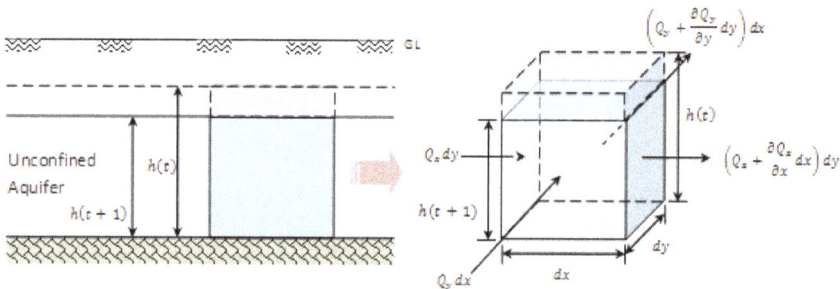

Elementary control volume for phreatic aquifer

Consider a phreatic aquifer as shown in the figure above. Let inflow per unit width of the aquifer in x direction is Q_x and that in y direction is Q_y.

Total inflow in x direction is

$Q_x dy$

Total outflow in x direction is

$$\left(Q_x + \frac{\partial Q_x}{\partial x} dx\right) dy$$

Net flow in x direction is,

$$-\frac{\partial Q_x}{\partial x} dxdy$$

Similarly the net flow in y direction is

$$-\frac{\partial Q_y}{\partial y} dxdy$$

Let $N(x, y, t)$ is the source or sink of the control volume per unit area.

Source or sink flow is therefore

$N(x, y, t)dxdy$

Total net flow of the control volume is

$$\left(-\frac{\partial Q_x}{\partial x} - \frac{\partial Q_y}{\partial y} + N(x,y,t)\right) dxdy$$

If S_y is the specific yield of the aquifer, then

$$S_y = \frac{\Delta V \omega}{A \Delta h}$$

$$\Rightarrow \Delta V \omega = S_y A \Delta h$$

Now change in storage in time dt can be written as,

$$S_y dxdy \frac{\partial h}{\partial t}$$

For maintaining continuity

Net flow in time dt = change in storage in time dt

$$\left(-\frac{\partial Q_x}{\partial x} - \frac{\partial Q_y}{\partial y} + N(x,y,t)\right) dxdy = S_y dxdy \frac{\partial h}{\partial t}$$

$$\Rightarrow -\frac{\partial Q_x}{\partial x} - \frac{\partial Q_y}{\partial y} + N(x,y,t) = S_y \frac{\partial h}{\partial t}$$

Now

$$Q_x = -K_x h \frac{\partial h}{\partial x}$$

$$Q_y = -K_y h \frac{\partial h}{\partial y}$$

Putting in

$$\Rightarrow \frac{\partial}{\partial x}\left(K_x h \frac{\partial h}{\partial x}\right) + \frac{\partial}{\partial y}\left(K_y h \frac{\partial h}{\partial y}\right) + N(x,y,t) = S_y \frac{\partial h}{\partial t}$$

In case of non-homogeneous isotropic aquifer

$$\frac{\partial}{\partial x}\left(Kh \frac{\partial h}{\partial x}\right) + \frac{\partial}{\partial y}\left(Kh \frac{\partial h}{\partial y}\right) + N = S_y \frac{\partial h}{\partial t}$$

In case of homogeneous isotropic aquifer

$$\frac{\partial}{\partial x}\left(h \frac{\partial h}{\partial x}\right) + \frac{\partial}{\partial y}\left(h \frac{\partial h}{\partial y}\right) + \frac{N}{K} = \frac{S_y}{K} \frac{\partial h}{\partial t}$$

This is the basic continuity equation for 2-D groundwater flow in a phreatic aquifer with a horizontal impervious base. This equation is also known as the Boussinesq equation. The equation derived above is a nonlinear one, because of the product term $h\partial h / \partial x$. The equation may be converted to a linear equation using the following techniques.

1. Assuming $T = \overline{T} + \Delta T$ where $\overline{T}$ is the average constant transmissivily of the aquifer and ΔT is the deviation from the average. The flow equation can be reduced to a linear equation by putting $\overline{T} = K\overline{h}$. In this case,

$$\overline{T}\left(\frac{\partial^2 h}{\partial x^2} + \frac{\partial^2 h}{\partial y^2}\right) + N = S_y \frac{\partial h}{\partial t}$$

2. In the second method the flow equation can be rewrite as

$$\frac{\partial^2 h^2}{\partial x^2} + \frac{\partial^2 h^2}{\partial y^2} + \frac{2N}{K} = \frac{S_y}{K} \frac{\partial h^2}{\partial t}$$

Putting, $K\overline{h} = T$

We have,

$$\frac{\partial^2 h^2}{\partial x^2} + \frac{\partial^2 h^2}{\partial y^2} + \frac{2N}{K} = \frac{S_y}{T}\frac{\partial h^2}{\partial t}$$

Two-dimensional Flow in Leaky Confined Aquifer

Consider the leaky confined aquifer as shown in figure below. The piezometric head in the main aquifer is (φ). The main aquifer is confined by the semipervious aquifer at top and bottom. The piezometric head in the top phreatic aquifer is (φ_1) and that for the bottom confined aquifer is (φ_2). The thickness of the main aquifer is B and the thickness of the top and bottom semi-pervious strata is B_1 and B_2 respectively. The hydraulic conductivity of the main aquifer is K and that for the top and bottom semipervious strata is K_1 and K_2 respectively.

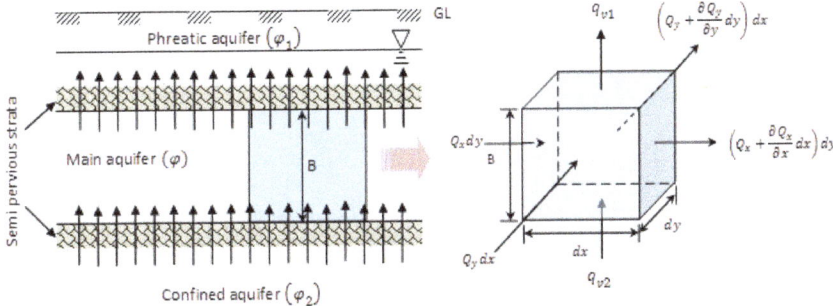

Elementary control volume for leaky confined aquifer

For the 2D control volume shown in figure above.

Let Q_x is the inflow per unit width in x direction

Q_y is the inflow per unit width in y direction

Inflow in x direction is

$Q_x dy$

Outflow in x direction is

$$\left(Q_x + \frac{\partial Q_x}{\partial_x}dx\right)dy$$

Net flow in x direction

$$= Q_x dy - \left(Q_x + \frac{\partial Q_x}{\partial_x}dx\right)dy$$

$$= -\frac{\partial Q_x}{\partial_x} dxdy$$

Similarly, net flow in y direction is

$$-\frac{\partial Q_y}{\partial_y} dxdy$$

The flow enter into the control volume from the bottom semi-pervious layer is

$$q_{v2}dxdy$$

The flow coming out from the control volume through the top semi-pervious layer is

$$q_{v1}dxdy$$

Total net flow of the control volume is

$$\left(-\frac{\partial Q_x}{\partial_x} - \frac{\partial Q_y}{\partial_y}\right)dxdy + \left(q_{v2} - q_{v1}\right)dxdy$$

Now, as we know that aquifer storativity

$$S_s = \frac{\Delta V_\omega}{A\Delta\varphi}$$

$$\Rightarrow \Delta V_\omega = S_s A\Delta\varphi$$

$$= S_s dxdy\Delta\varphi$$

Change in storage in time dt is

$$S_s \frac{\partial\varphi}{\partial t} dxdy$$

Total net flow = change in storage in time dt

$$\left(-\frac{\partial Q_x}{\partial_x} - \frac{\partial Q_y}{\partial_y}\right)dxdy + \left(q_{v2} - q_{v1}\right)dxdy = S_s \frac{\partial\varphi}{\partial t} dxdy$$

$$\Rightarrow -\left(\frac{\partial Q_x}{\partial_x} + \frac{\partial Q_y}{\partial_y}\right) + \left(q_{v2} - q_{v1}\right) = S_s \frac{\partial\varphi}{\partial t}$$

Now putting,

$$Q_x = -T_x \frac{\partial \varphi}{\partial_x} \text{ and } Q_y = -T_y \frac{\partial \varphi}{\partial_y}$$

$$q_{v1} = K_1 \frac{\varphi - \varphi_1}{B_1} \text{ and } q_{v2} = K_2 \frac{\varphi_2 - \varphi}{B_2}$$

We have,

$$\frac{\partial}{\partial_x} \left(T_x \frac{\partial \varphi}{\partial_x} \right) + \frac{\partial}{\partial_y} \left(T_y \frac{\partial \varphi}{\partial_y} \right) + \left(K_2 \frac{\varphi_2 - \varphi}{B_2} - K_1 \frac{\varphi - \varphi_1}{B_1} \right) = S_s \frac{\partial \varphi}{\partial t}$$

$$\frac{\partial}{\partial_x} \left(T_x \frac{\partial \varphi}{\partial_x} \right) + \frac{\partial}{\partial_y} \left(T_y \frac{\partial \varphi}{\partial_y} \right) + \left(\frac{\varphi_2 - \varphi}{B_2 / K_2} - \frac{\varphi - \varphi_1}{B_1 / K_1} \right) = S_s \frac{\partial \varphi}{\partial t}$$

Putting $B_2 / K_2 = \sigma_2$ and $B_1 / K_1 = \sigma_1$, we have

$$\frac{\partial}{\partial_x} \left(T_x \frac{\partial \varphi}{\partial_x} \right) + \frac{\partial}{\partial_y} \left(T_y \frac{\partial \varphi}{\partial_y} \right) + \left(\frac{\varphi_2 - \varphi}{\sigma_2} - \frac{\varphi - \varphi_1}{\sigma_1} \right) = S_s \frac{\partial \varphi}{\partial t}$$

This is the flow equation for anisotropic non-homogeneous leaky confined aquifer for unsteady condition.

In case of homogeneous aquifer, the equation becomes

$$T_x \frac{\partial^2 \varphi}{\partial_{x^2}} + T_y \frac{\partial^2 \varphi}{\partial_{y^2}} + \left(\frac{\varphi_2 - \varphi}{\sigma_2} - \frac{\varphi - \varphi_1}{\sigma_1} \right) = S_s \frac{\partial \varphi}{\partial t}$$

For homogeneous and isotropic condition,

$$\frac{\partial^2 \varphi}{\partial_{x^2}} + \frac{\partial^2 \varphi}{\partial_{y^2}} + \left(\frac{\varphi_2 - \varphi}{T \sigma_2} - \frac{\varphi - \varphi_1}{T \sigma_1} \right) = \frac{S_s}{T} \frac{\partial \varphi}{\partial t}$$

Putting $T \sigma_1 = \lambda_1$ and $T \sigma_2 = \lambda_2$

$$\frac{\partial^2 \varphi}{\partial_{x^2}} + \frac{\partial^2 \varphi}{\partial_{y^2}} + \left(\frac{\varphi_2 - \varphi}{\lambda_2} - \frac{\varphi - \varphi_1}{\lambda_1} \right) = \frac{S_s}{T} \frac{\partial \varphi}{\partial t}$$

The λ_1 and λ_2 are known as leakage factor of the semi-pervious layer.

If a source $N(x,y,t)$ is present, the equation becomes

$$\frac{\partial^2 \varphi}{\partial x^2} + \frac{\partial^2 \varphi}{\partial y^2} + \left(\frac{\varphi_2 - \varphi}{\lambda_2} - \frac{\varphi - \varphi_1}{\lambda_1} \right) + N(x,y,t) = \frac{S_s}{T} \frac{\partial \varphi}{\partial t}$$

Two-dimensional Flow in Leaky Unconfined Aquifer

Consider the leaky unconfined aquifer as shown in figure below. The bottom of the unconfined aquifer has a semi pervious layer. The piezometric head in the bottom confined aquifer is (φ). The thickness of the bottom semi-pervious strata is B. The hydraulic conductivity of the main aquifer is K and that bottom semi-pervious strata is K₁.

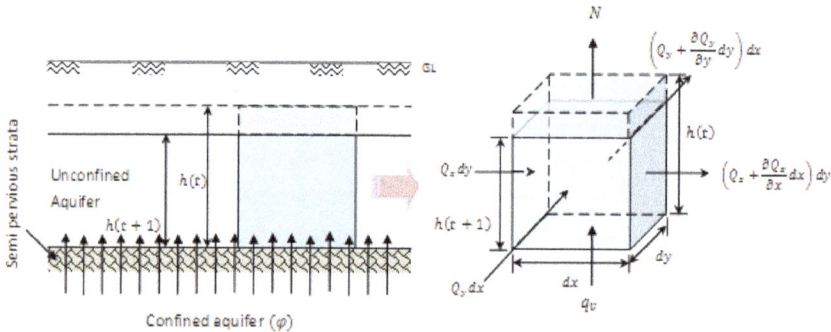

Elementary control volume for leaky unconfined aquifer

For the 2D control volume shown in figure above.

Let inflow per unit width of the aquifer in x direction is Q_x and that in y direction is Q_y.

Total inflow in x direction is

$$Q_x dy$$

Total outflow in x direction is

$$\left(Q_x + \frac{\partial Q_x}{\partial x} dx \right) dy$$

Net flow in x direction is,

$$-\frac{\partial Q_x}{\partial x} dx dy$$

Similarly the net flow in y direction is

$$-\frac{\partial Q_y}{\partial_y}\,dxdy$$

The flow enter into the control volume from the bottom semi-pervious layer is

$$q_v dxdy$$

Let $N(x,y,t)$ is the source or sink of the control volume per unit area.

Source or sink flow is therefore

$$N(x,y,t)\,dxdy$$

Total net flow of the control volume is

$$\left(-\frac{\partial Q_x}{\partial_x}-\frac{\partial Q_y}{\partial_y}+q_v+N(x,y,t)\right)dxdy$$

If S_y is the specific yield of the aquifer, then

$$S_y=\frac{\Delta V_\omega}{A\Delta h}$$

$$\Rightarrow \Delta V_\omega = S_y A\Delta h$$

Now change in storage in time dt can be written as,

$$S_y dxdy\,\frac{\partial h}{\partial t}$$

For maintaining continuity

Net flow in time dt = change in storage in time dt

$$\left(-\frac{\partial Q_x}{\partial_x}-\frac{\partial Q_y}{\partial_y}+q_v+N(x,y,t)\right)dxdy=S_y dxdy\,\frac{\partial h}{\partial t}$$

$$\Rightarrow -\frac{\partial Q_x}{\partial_x}-\frac{\partial Q_y}{\partial_y}+q_v+N(x,y,t)=S_y\,\frac{\partial h}{\partial t}$$

Now

$$Q_x = -K_x h \frac{\partial h}{\partial x}$$

$$Q_y = -K_y h \frac{\partial h}{\partial y}$$

$$q_{v1} = K_1 \frac{\varphi - h}{B}$$

Putting in

$$\frac{\partial}{\partial x}\left(K_x h \frac{\partial h}{\partial x}\right) + \frac{\partial}{\partial y}\left(K_y h \frac{\partial h}{\partial y}\right) + K\frac{-h}{B} + N(x, y, t) = S_y \frac{\partial h}{\partial t}$$

$$\Rightarrow \frac{\partial}{\partial x}\left(K_x h \frac{\partial h}{\partial x}\right) + \frac{\partial}{\partial y}\left(K_y h \frac{\partial h}{\partial y}\right) + \frac{\varphi - h}{B / K_1} + N(x, y, t) = S_y \frac{\partial h}{\partial t}$$

Putting, $\sigma = B / K_1$

$$\Rightarrow \frac{\partial}{\partial x}\left(K_x h \frac{\partial h}{\partial x}\right) + \frac{\partial}{\partial y}\left(K_y h \frac{\partial h}{\partial y}\right) + \frac{\varphi - h}{\sigma} + N(x, y, t) = S_y \frac{\partial h}{\partial t}$$

In case of non-homogeneous isotropic aquifer

$$\frac{\partial}{\partial x}\left(Kh \frac{\partial h}{\partial x}\right) + \frac{\partial}{\partial y}\left(Kh \frac{\partial h}{\partial y}\right) + \frac{\varphi - h}{\sigma} + N(x, y, t) = S_y \frac{\partial h}{\partial t}$$

Three Dimensional Flow

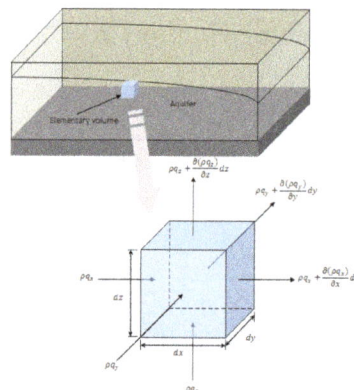

Elementary control volume

Consider the control volume as shown in the figure above. The dimension of the rectangular parallelepiped box is dx, dy and dz. Let q_x, q_y and q_z are the volumetric flow per unit area entering in to the control volume through $(y$ - $z)$, $(x$ - $z)$ and $(x$ - $y)$ faces respectively.

The mass flux inflow along x direction is $\rho q x dy dz$.

Outflow in the x - direction will be $\left(\rho q_x + \dfrac{\partial(\rho q x)}{\partial x} dx \right) dy dz$

The excess of inflow over outflow of mass during time interval can be expressed as,

$$\left[\rho q_x dy dz - \left(\rho q_x + \frac{\partial(\rho q_x)}{\partial x} dx \right) dy dz \right]$$

$$= -\frac{\partial(\rho q_x)}{\partial x} dx dy dz$$

Similarly in y and z direction, we have

$$-\frac{\partial(\rho q_y)}{\partial y} dx dy dz \text{ and } \frac{\partial(\rho q_z)}{\partial z} dx dy dz$$

Total excess of mass inflow over outflow can be expressed as:

$$-\left[\frac{\partial(\rho q_x)}{\partial x} + \frac{\partial(\rho q_y)}{\partial y} + \frac{\partial(\rho q_z)}{\partial z} \right] dx dy dz$$

Now, as per the principle of mass conservation, the excess of mass must be equal to the change in mass during dt. Change in storage can be written as ∂v_w.

Now,

$$\partial v_w = S_0 V_T (\partial \varphi)$$

Where S_o is the specific storage,

V_T is the total volume of the porous matrix

Then,

$$\partial v_w = S_0 dx dy dz\, \partial \varphi$$

Now rate of change of storage can be written as:

$$\frac{\partial v_w}{\partial t} = \rho S_0 dx dy dz \frac{\partial \varphi}{\partial t}$$

Putting this in the flow equation, one can have

$$-\frac{\partial(\rho q_x)}{\partial x} - \frac{\partial(\rho q_y)}{\partial y} - \frac{\partial(\rho q_z)}{\partial z} = S_0 \rho \frac{\partial \varphi}{\partial z}$$

Considering water is incompressible (ρ = constant)

$$-\frac{\partial q_x}{\partial x} - \frac{\partial q_y}{\partial y} - \frac{\partial q_z}{\partial z} = S_0 \frac{\partial \varphi}{\partial t}$$

$Or, -\nabla.q = S_0 \dfrac{\partial \varphi}{\partial t}$

Now, as per Darcy's law q_x, q_y and q_z can be written as

$$q_x = -K_x \frac{\partial \varphi}{\partial x}$$

$$q_y = -K_y \frac{\partial \varphi}{\partial y}$$

$$q_z = -K_z \frac{\partial \varphi}{\partial z}$$

Putting in flow equation

$$-\frac{\partial}{\partial x}\left(-K_x \frac{\partial \varphi}{\partial x}\right) - \frac{\partial}{\partial y}\left(-K_y \frac{\partial \varphi}{\partial y}\right) - \frac{\partial}{\partial z}\left(-K_z \frac{\partial \varphi}{\partial z}\right) = S_0 \frac{\partial \varphi}{\partial t}$$

$$\Rightarrow \frac{\partial}{\partial x}\left(K_x \frac{\partial \varphi}{\partial x}\right) + \frac{\partial}{\partial y}\left(K_y \frac{\partial \varphi}{\partial y}\right) + \frac{\partial}{\partial z}\left(K_z \frac{\partial \varphi}{\partial z}\right) = S_0 \frac{\partial \varphi}{\partial t}$$

$$\Rightarrow \nabla.K\nabla\varphi = S_0 \frac{\partial \varphi}{\partial t}$$

In case of homogeneous aquifer,

$$K_x \frac{\partial^2 \varphi}{\partial x^2} + K_y \frac{\partial^2 \varphi}{\partial y^2} + K_z \frac{\partial^2 \varphi}{\partial z^2} = S_0 \frac{\partial \varphi}{\partial t}$$

In case of homogeneous isotropic aquifer ($K_x = K_y = K_z = K$)

$$\frac{\partial^2 \varphi}{\partial x^2} + \frac{\partial^2 \varphi}{\partial y^2} + \frac{\partial^2 \varphi}{\partial z^2} = \frac{S_0}{K} \frac{\partial \varphi}{\partial t}$$

$Or, \nabla^2 \varphi = \dfrac{S_0}{K} \dfrac{\partial \varphi}{\partial t}$

For steady state condition $\left(\partial \varphi / \partial t = 0 \right)$ the equation becomes

$$\frac{\partial^2 \varphi}{\partial x^2} + \frac{\partial^2 \varphi}{\partial y^2} + \frac{\partial^2 \varphi}{\partial z^2} = 0$$

For 2-D steady state condition, it will be

$$\frac{\partial^2 \varphi}{\partial x^2} + \frac{\partial^2 \varphi}{\partial y^2} = 0$$

If we have distributed sources and sinks of $N(x,y,z,t)$ in the aquifer, the flow equation becomes

$$\frac{\partial}{\partial x}\left(K_x \frac{\partial \varphi}{\partial x} \right) + \frac{\partial}{\partial y}\left(K_y \frac{\partial \varphi}{\partial y} \right) + \frac{\partial}{\partial z}\left(K_z \frac{\partial \varphi}{\partial z} \right) + N(x, y, z, t) = S_0 \frac{\partial \varphi}{\partial t}$$

Boundary and Initial Conditions

For solving the steady state flow equation as derived above, appropriate boundary conditions are needed. It is one of the required components of the mathematical model. On the other hand, for solving transient flow equation, appropriate initial condition is also required. Boundary conditions are generally three types. They are Dirichlet boundary condition, Neumann boundary condition and mixed boundary condition.

Dirchlet Boundary Condition

In case of Dirichlet boundary condition, prescribe value of the variable $h(x,y,z,t)$ is specified at the boundary of the problem domain. This is also known as type I boundary condition. The head may be constant or may vary in space or in time.

Neumann Boundary Condition

In case of Neumann boundary condition, the gradient of the variable ($\partial h / \partial n$) is specified at the boundary of the problem domain. Here, n is the direction, x, y, and z. One of the most frequently use Neumann boundary condition is the no flow boundary condition, i.e. $\partial h / \partial n = 0$ at the boundary.

As discussed, in case of Neumann boundary condition, we have

$$\frac{\partial h}{\partial n} = C_1$$

Where, C_1 is constant

$$\Rightarrow -K_n \frac{\partial h}{\partial n} = -K_n C_1 = q_n$$

Where q_n is the Darcy ›s flux in the nth direction. As such in case of Neumann boundary condition, we can also specify the Darcy's flux at the boundary instead of the gradient of the variable. The Neumann boundary condition is also knows as Type II boundary condition.

Mixed Boundary Condition

We can also specify a mixed boundary condition in the form given below.

$$ah + b\frac{\partial h}{\partial n} = \text{Constant}$$

This is also known as Type III boundary condition and is the linear combination of Type I and Type II boundary condition.

Initial Condition

For time dependent problem, an initial condition for the head field $h(x,y,z,t = 0)$ has to be specified. This is known as initial condition.

Figure (a) shows the flow domain between two observation wells in case of a confined aquifer. The lines CF and DE represent equpotential lines of potential φ_0 and φ_L respectively. Therefore on these two boundaries, Dirchlet boundary is to be applied. On the other hand at top and bottom of the surfaces of the aquifer are impervious. The lines CD and EF represent the no flow boundary. As such the Neumann boundary condition is to be applied on these two sides. Figure (b) shows the aquifer with boundary conditions.

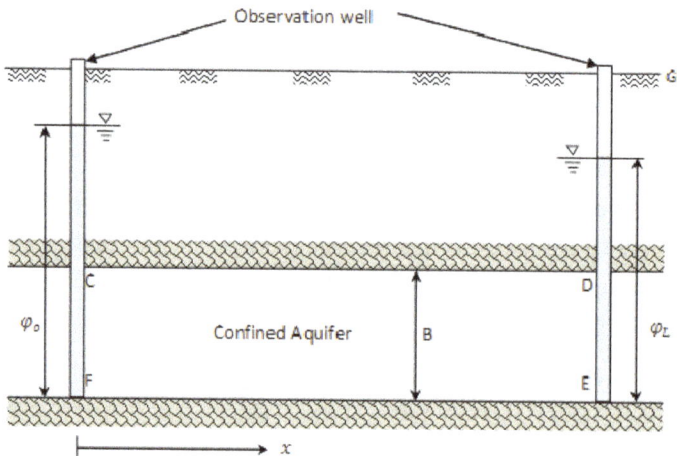

(a) Flow domain between observation wells

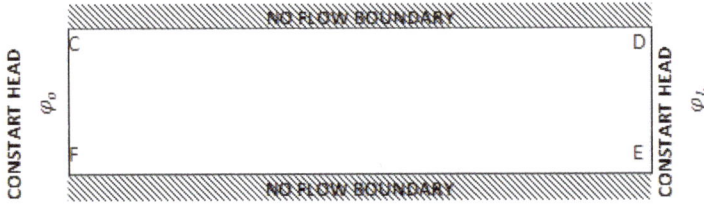

(b) Confined aquifer with boundary condition

References

- Carrigy, N.; Pant, L. M.; Mitra, S. K.; Secanell, M. (2013). "Knudsen diffusivity and permeability of pemfc microporous coated gas diffusion layers for different polytetrafluoroethylene loadings". Journal of the Electrochemical Society. 160: F81–89. doi:10.1149/2.036302jes

- Whitaker, S. (1986). "Flow in porous media I: A theoretical derivation of Darcy's law". Transport in Porous Media. 1: 3–25. doi:10.1007/BF01036523

- Peter C. Lichtner, Carl I. Steefel, Eric H. Oelkers, Reactive Transport in Porous Media, Mineralogical Society of America, 1996, ISBN 0-939950-42-1, p. 5

- Pant, L. M.; Mitra, S. K.; Secanell, M. (2012). "Absolute permeability and Knudsen diffusivity measurements in PEMFC gas diffusion layers and micro porous layers". Journal of Power Sources. 206: 153–160. doi:10.1016/j.jpowsour.2012.01.099

- Brinkman, H. C. (1949). "A calculation of the viscous force exerted by a flowing fluid on a dense swarm of particles". Applied Scientific Research. 1: 27–34. doi:10.1007/BF02120313

- Kerkhof, P. (1996). "A modified Maxwell–Stefan model for transport through inert membranes: The binary friction model". Chemical Engineering Journal and the Biochemical Engineering Journal. 64: 319–343. doi:10.1016/S0923-0467(96)03134-X

- Jin, Y.; Uth, M.-F.; Kuznetsov, A. V.; Herwig, H. (2 February 2015). "Numerical investigation of the possibility of macroscopic turbulence in porous media: a direct numerical simulation study". Journal of Fluid Mechanics. 766: 76–103. Bibcode:2015JFM...766...76J. doi:10.1017/jfm.2015.9

Aquifer Water Flow: Problems and Solutions

Water equilibrium can be achieved in confined and unconfined aquifer. If the pumping of water into a well is constant, the simultaneous refilling of the well causes water flow to be radially symmetric. The chapter closely examines the key concepts of water equilibrium to provide an extensive understanding of the subject.

Water Flow

The flow towards a well, situated in homogeneous and isotropic confined or unconfined aquifer is radially symmetric. Figure below shows the cone of depression caused due to constant pumping through a single well situated at (0,0) in a confined aquifer. Figure below shows the cone of impression caused due to constant recharge through the well. In case of homogeneous and isotropic medium, the cone of depression or cone of impression is radially symmetrical. The governing equation derived earlier in Cartesian coordinate system for confined and unconfined aquifer can also be derived for radial flow in an aquifer. In this lesson, we will derive the governing flow equation for confined and unconfined aquifer in polar coordinate system. The main objective of this conversion is to make the 2D flow problem a 1D flow problem. The resulting 1D problem will be simpler to solve.

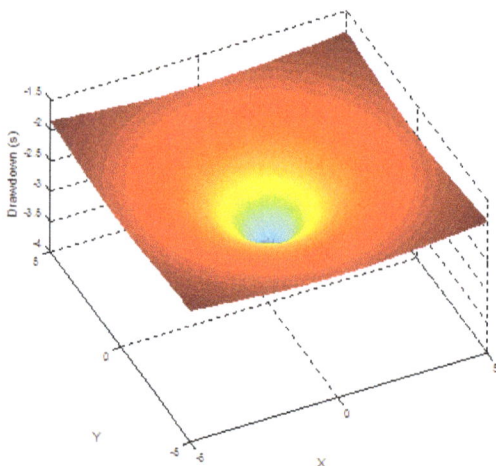

Cone of depression

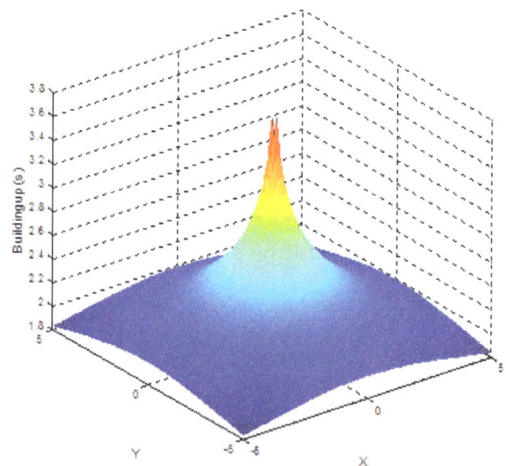

Cone of impression

Confined Aquifer

- Radial flow to a well

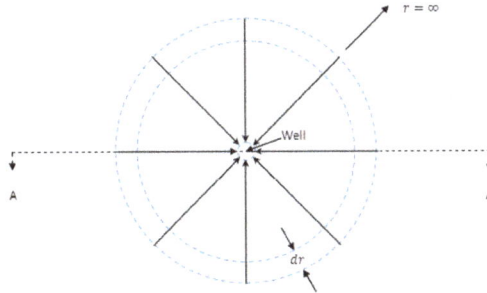

- Section A-A in case of confined aquifer

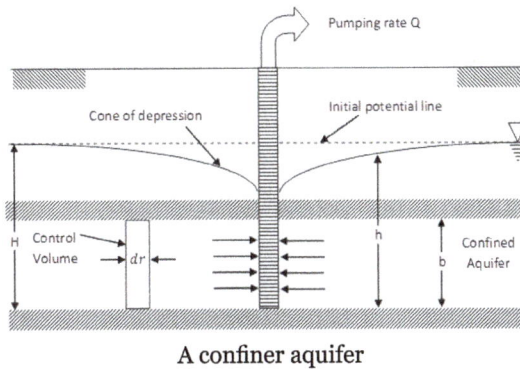

A confiner aquifer

Let us consider a case of radial flow to a single well in a confined aquifer. The figure above shows the radial flow towards a well and a control volume of thickness dr. The figure above shows the vertical section AA of the aquifer along with cone of depression. The aquifer is homogeneous and isotropic and have constant thickness of b. The hydraulic conductivity of the aquifer is K. The pumping rate (Q) of the aquifer is constant and the well diameter is infinitesimally small. The well is fully penetrated into the entire thickness of the confined aquifer. This is necessary to make the flow essentially horizontal. The potential head in the aquifer prior to pumping is uniform throughout the aquifer.

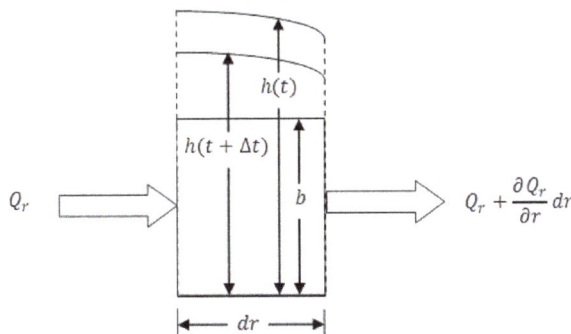

Control volume in case of confined aquifer

Consider the control volume shown in figure above:

The inflow to the control volume is Q_r

The outflow from the control volume is $Q_r + \frac{\partial Q_r}{\partial r} dr$

The net inflow to the control volume is $Q_r - \left(Q_r + \frac{\partial Q_r}{\partial r} dr \right) = -\frac{\partial Q_r}{\partial r} dr$

Applying principle of mass conservation on the control volume

Inflow - outflow = Time rate of change in volumetric storage

Time rate of change in volumetric storage $= \frac{\partial V}{\partial t} = V \left(\frac{\partial V}{V \partial h} \right) \frac{\partial h}{\partial t}$

$$\frac{\partial V}{\partial t} = V S_0 \frac{\partial h}{\partial t}$$

where S_o is the specific storage $= \frac{\partial V}{v \partial h}$

Replacing V by $2\pi r dr b$, we have

$$\frac{\partial V}{\partial t} = V S_0 \frac{\partial h}{\partial t} = 2\pi r dr b s_0 \frac{\partial h}{\partial t}$$

$$= 2\pi r dr b \frac{S_s}{b} \frac{\partial h}{\partial t} = 2\pi r dr b S_s \frac{\partial h}{\partial t}$$

Where S_s is the aquifer storativity which is equal to S_o / b

Adding the above equations, we have

$$-\frac{\partial Q_r}{\partial_r} dr = 2\pi r dr b S_s \frac{\partial h}{\partial t}$$

As per Darcy's law

$$Q_r = -KA \frac{\partial h}{\partial t} = -K 2\pi r b \frac{\partial h}{\partial t} = -2\pi r T \frac{\partial h}{\partial t} \quad [Putting\ T = Kb]$$

Putting in the above equation,

$$\frac{\partial}{\partial r} \left(2\pi r T \frac{\partial h}{\partial t} \right) dr = 2\pi r dr b S_s \frac{\partial h}{\partial t}$$

Simplifying,

$$\frac{1}{r} \frac{\partial}{\partial r} \left(r \frac{\partial h}{\partial r} \right) = \frac{S_s}{T} \frac{\partial h}{\partial t}$$

$$\frac{\partial^2 h}{\partial r^2} + \frac{1}{r}\frac{\partial h}{\partial r} = \frac{S_S}{T}\frac{\partial h}{\partial t}$$

This is the flow equation for radial flow into a well for confined homogeneous and isotropic aquifer.

In case of steady state condition, the governing equation becomes,

$$\frac{\partial^2 h}{\partial r^2} + \frac{1}{r}\frac{\partial h}{\partial r} = 0$$

Unconfined Aquifer

(a) Radial flow to a well

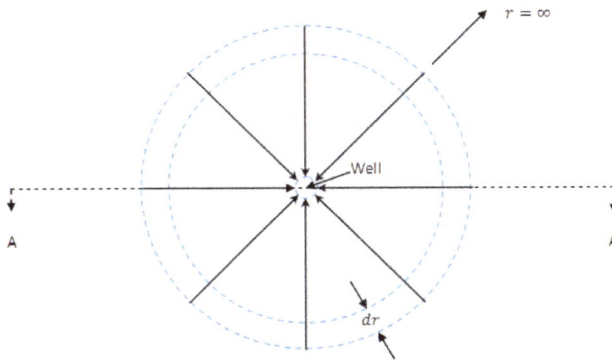

(b) Section A-A in case of unconfined aquifer

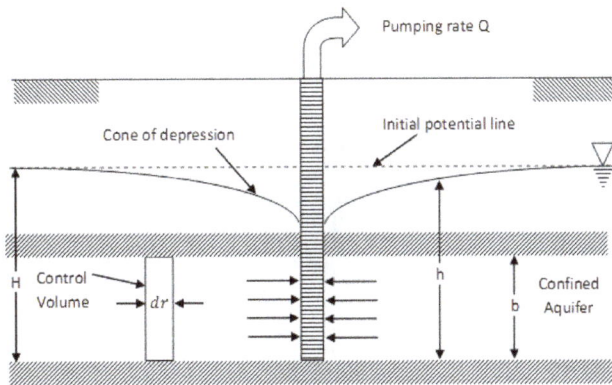

An unconfined aquifer

Let us consider a case of radial flow to a single well. The unconfined aquifer is homogeneous and isotropic. The hydraulic conductivity of the aquifer is K. The pumping rate (Q) of the aquifer is constant and the well diameter is infinitesimally small. The well is fully penetrated into the aquifer and hydraulic head in the aquifer prior to pumping is uniform throughout the aquifer.

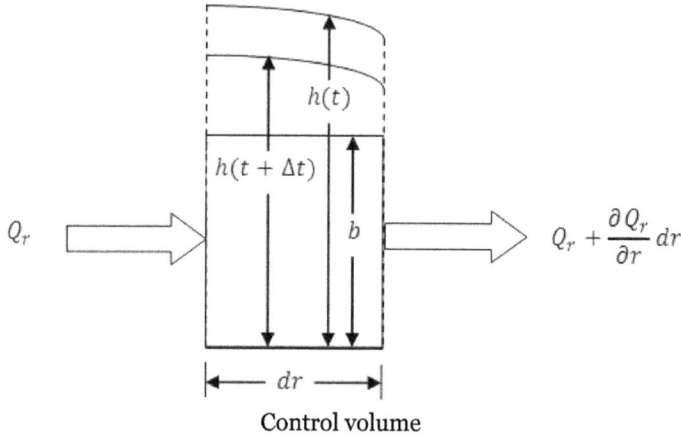

Control volume

For the control volume shown in figure above,

The inflow to the system is Q_r

The outflow from the system is $Qr + \dfrac{\partial Qr}{\partial r} dr$

The net inflow to the system is $Qr - \left(Qr + \dfrac{\partial Qr}{\partial r} dr \right) = -\dfrac{\partial Qr}{\partial r} dr$

Applying principle of mass conservation on the control volume

Inflow - outflow = Time rate of change in volumetric storage

Time rate of change in volumetric storage $= \dfrac{\partial V}{\partial t} = V \left(\dfrac{\partial V}{V \partial t} \right) \dfrac{\partial h}{\partial t}$

$$\dfrac{\partial V}{\partial t} = V S_0 \dfrac{\partial h}{\partial t}$$

where S_o is the specific storage $= \dfrac{\partial V}{V \partial h}$

Replacing V by $2\pi r dr h$, we have

$$\dfrac{\partial V}{\partial t} = V S_0 \dfrac{\partial h}{\partial t} = 2\pi r dr h S_0 \dfrac{\partial h}{\partial t}$$

$$= 2\pi r dr h \dfrac{S_y}{h} \dfrac{\partial h}{\partial t} = 2\pi r dr S_y \dfrac{\partial h}{\partial t}$$

Where S_y is the specific yield which is equal to S_o / h.

Now substituting the above equations, we have

$$-\dfrac{\partial Qr}{\partial r} dr = 2\pi r dr S_y \dfrac{\partial h}{\partial t}$$

As per Darcy's law

$$Qr = -KA\frac{\partial h}{\partial t} = -K2\pi rh\frac{\partial h}{\partial t}$$

Putting in equation above

$$\frac{\partial}{\partial r}\left(K2\pi rh\frac{\partial h}{\partial t}\right)dr = 2\pi rdrS_y\frac{\partial h}{\partial t}$$

Simplifying, $\dfrac{1}{r}\dfrac{\partial}{\partial r}\left(rh\dfrac{\partial h}{\partial r}\right) = \dfrac{S_y}{K}\dfrac{\partial h}{\partial t}$

$$\Rightarrow \frac{1}{r}\frac{\partial}{\partial r}\left(\frac{r}{2}\frac{\partial h^2}{\partial r}\right) = \frac{S_y}{K}\frac{\partial h}{\partial t}$$

$$\Rightarrow \frac{1}{2r}\frac{\partial r}{\partial r}\frac{\partial h^2}{\partial r} + \frac{1}{r}\frac{r}{2}\frac{\partial^2 h^2}{\partial r^2} = \frac{S_y}{K}\frac{\partial h}{\partial t}$$

$$\Rightarrow \frac{1}{r}\frac{\partial h^2}{\partial r} + \frac{\partial^2 h^2}{\partial r^2} = \frac{2S_y}{K}\frac{\partial h}{\partial t}$$

This is the flow equation for radial flow into a well for unconfined homogeneous isotropic aquifer.

In case of steady state condition, the governing equation becomes,

$$\frac{1}{r}\frac{\partial h^2}{\partial r} + \frac{\partial^2 h^2}{\partial r^2} = 0$$

Or, $\dfrac{\partial}{\partial r}\left(rh\dfrac{\partial h}{\partial t}\right) = 0$

Fluid Dynamics

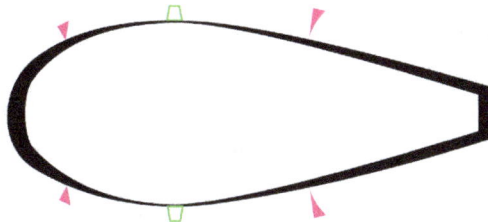

Typical aerodynamic teardrop shape, assuming a viscous medium passing from left to right, the diagram shows the pressure distribution as the thickness of the black line and shows the velocity in the boundary layer as the violet triangles. The green vortex generators prompt the transition to turbulent flow and prevent back-flow also called flow separation from the high pressure region in the back. The surface in front is as smooth as possible or even employs shark-like skin, as any turbulence here increases the energy of the airflow. The truncation on the right, known as a Kammback, also prevents backflow from the high pressure region in the back across the spoilers to the convergent part.

In physics and engineering, fluid dynamics is a subdiscipline of fluid mechanics that describes the flow of fluids (liquids and gases). It has several subdisciplines, including aerodynamics (the study of air and other gases in motion) and hydrodynamics (the study of liquids in motion). Fluid dynamics has a wide range of applications, including calculating forces and moments on aircraft, determining the mass flow rate of petroleum through pipelines, predicting weather patterns, understanding nebulae in interstellar space and modelling fission weapon detonation.

Fluid dynamics offers a systematic structure—which underlies these practical disciplines—that embraces empirical and semi-empirical laws derived from flow measurement and used to solve practical problems. The solution to a fluid dynamics problem typically involves the calculation of various properties of the fluid, such as flow velocity, pressure, density, and temperature, as functions of space and time.

Before the twentieth century, *hydrodynamics* was synonymous with fluid dynamics. This is still reflected in names of some fluid dynamics topics, like magnetohydrodynamics and hydrodynamic stability, both of which can also be applied to gases.

Equations of Fluid Dynamics

The foundational axioms of fluid dynamics are the conservation laws, specifically, conservation of mass, conservation of linear momentum (also known as Newton's Second Law of Motion), and conservation of energy (also known as First Law of Thermodynamics). These are based on classical mechanics and are modified in quantum mechanics and general relativity. They are expressed using the Reynolds transport theorem.

In addition to the above, fluids are assumed to obey the continuum assumption. Fluids are composed of molecules that collide with one another and solid objects. However, the continuum assumption assumes that fluids are continuous, rather than discrete. Consequently, it is assumed that properties such as density, pressure, temperature, and flow velocity are well-defined at infinitesimally small points in space and vary continuously from one point to another. The fact that the fluid is made up of discrete molecules is ignored.

For fluids that are sufficiently dense to be a continuum, do not contain ionized species, and have flow velocities small in relation to the speed of light, the momentum equations for Newtonian fluids are the Navier–Stokes equations—which is a non-linear set of differential equations that describes the flow of a fluid whose stress depends linearly on flow velocity gradients and pressure. The unsimplified equations do not have a general closed-form solution, so they are primarily of use in Computational Fluid Dynamics. The equations can be simplified in a number of ways, all of which make them easier to solve. Some of the simplifications allow some simple fluid dynamics problems to be solved in closed form.

In addition to the mass, momentum, and energy conservation equations, a thermodynamic equation of state that gives the pressure as a function of other thermodynamic variables is required to completely describe the problem. An example of this would be the perfect gas equation of state:

$$p = \frac{\rho R_u T}{M}$$

where p is pressure, ρ is density, T the absolute temperature, while R_u is the gas constant and M is molar mass for a particular gas.

Conservation Laws

Three conservation laws are used to solve fluid dynamics problems, and may be written in integral or differential form. The conservation laws may be applied to a region of the flow called a *control volume*. A control volume is a discrete volume in space through which fluid is assumed to flow. The integral formulations of the conservation laws are used to describe the change of mass, momentum, or energy within the control volume. Differential formulations of the conservation laws apply Stokes' theorem to yield an expression which may be interpreted as the integral form of the law applied to an infinitesimally small volume (at a point) within the flow.

- Mass continuity (conservation of mass): The rate of change of fluid mass inside a control volume must be equal to the net rate of fluid flow into the volume. Physically, this statement requires that mass is neither created nor destroyed in the control volume, and can be translated into the integral form of the continuity equation:

$$\frac{\partial}{\partial t} \iiint_V \rho dV = - \oiint S\, \rho \mathbf{u} \cdot d\mathbf{S}$$

Above, ρ is the fluid density, u is the flow velocity vector, and t is time. The left-hand side of the above expression is the rate of increase of mass within the volume and contains a triple integral over the control volume, whereas the right-hand side contains an integration over the surface of the control volume of mass convected into the system. Mass flow into the system is accounted as positive, and since the normal vector to the surface is opposite the sense of flow into the system the term is negated. The differential form of the continuity equation is, by the divergence theorem:

$$\frac{\partial \rho}{\partial t} + \nabla \cdot (\rho \mathbf{u}) = 0$$

- Conservation of momentum: Newton's second law of motion applied to a control volume, is a statement that any change in momentum of the fluid within that control volume will be due to the net flow of momentum into the volume and the action of external forces acting on the fluid within the volume.

$$\frac{\partial}{\partial t} \iiint_V \rho \mathbf{u} dV = -\oiint_S (\rho \mathbf{u} \cdot d\mathbf{S}) \mathbf{u} - \oiint_S p d\mathbf{S} + \iiint_V \rho \mathbf{f}_{body} dV + \mathbf{F}_{surf}$$

In the above integral formulation of this equation, the term on the left is the net change of momentum within the volume. The first term on the right is the net rate at which momentum is convected into the volume. The second term on the right is the force due to pressure on the volume's surfaces. The first two terms on the right are negated since momentum entering the system is accounted as positive, and the normal is opposite the direction of the velocity $\mathbf{u}$ and pressure forces. The third term on the right is the net acceleration of the mass within the volume due to any body forces (here represented by f_{body}). Surface forces, such as viscous forces, are represented by $\mathbf{F}_{surf}$, the net force due to shear forces acting on the volume surface.

The following is the differential form of the momentum conservation equation. Here, the volume is reduced to an infinitesimally small point, and both surface and body forces are accounted for in one total force, F. For example, F may be expanded into an expression for the frictional and gravitational forces acting at a point in a flow.

$$\frac{D\mathbf{u}}{Dt} = \mathbf{F} - \frac{\nabla p}{\rho}$$

In aerodynamics, air is assumed to be a Newtonian fluid, which posits a linear relationship between the shear stress (due to internal friction forces) and the rate of strain of the fluid. The equation above is a vector equation in a three-dimensional flow, but it can be expressed as three scalar equations in three coordinate directions. The conservation of momentum equations for the compressible, viscous flow case are called the Navier–Stokes equations.

- Conservation of energy: Although energy can be converted from one form to another, the total energy in a closed system remains constant.

$$\rho \frac{Dh}{Dt} = \frac{Dp}{Dt} + \nabla \cdot (k \nabla T) + \Phi$$

Above, h is enthalpy, k is the thermal conductivity of the fluid, T is temperature, and Φ is the viscous dissipation function. The viscous dissipation function governs the rate at which mechanical energy of the flow is converted to heat. The second law of thermodynamics requires that the dissipation term is always positive: viscosity cannot create energy within the control volume. The expression on the left side is a material derivative.

Compressible vs Incompressible Flow

All fluids are compressible to some extent; that is, changes in pressure or temperature cause changes in density. However, in many situations the changes in pressure and temperature are sufficiently small that the changes in density are negligible. In this case the flow can be modelled as an incompressible flow. Otherwise the more general compressible flow equations must be used.

Mathematically, incompressibility is expressed by saying that the density ρ of a fluid parcel does not change as it moves in the flow field, i.e.,

$$\frac{D\rho}{Dt} = 0,$$

where D/Dt is the material derivative, which is the sum of local and convective derivatives. This additional constraint simplifies the governing equations, especially in the case when the fluid has a uniform density.

For flow of gases, to determine whether to use compressible or incompressible fluid dynamics, the Mach number of the flow is evaluated. As a rough guide, compressible effects can be ignored at Mach numbers below approximately 0.3. For liquids, whether the incompressible assumption is valid depends on the fluid properties (specifically the critical pressure and temperature of the fluid) and the flow conditions (how close to the critical pressure the actual flow pressure becomes). Acoustic problems always require allowing compressibility, since sound waves are compression waves involving changes in pressure and density of the medium through which they propagate.

Newtonian vs Non-Newtonian Fluids

All fluids are viscous, meaning that they exert some resistance to deformation: neighbouring parcels of fluid moving at different velocities exert viscous forces on each other. The velocity gradient is referred to as a strain rate; it has dimensions T^{-1}. Isaac Newton showed that for many familiar fluids such as water and air, the stress due to these viscous forces is linearly related to the strain rate. Such fluids are called Newtonian fluids. The coefficient of proportionality is called the fluid's viscosity; for Newtonian fluids, it is a fluid property that is independent of the strain rate.

Potential flow around an airfoil

Non-Newtonian fluids have a more complicated, non-linear stress-strain behaviour. The sub-discipline of rheology describes the stress-strain behaviours of such fluids, which include emulsions and slurries, some viscoelastic materials such as blood and some polymers, and *sticky liquids* such as latex, honey and lubricants.

Inviscid vs Viscous vs Stokes Flow

The dynamic of fluid parcels is described with the help of Newton's second law. An accelerating parcel of fluid is subject to inertial effects.

The Reynolds number is a dimensionless quantity which characterises the magnitude of inertial effects compared to the magnitude of viscous effects. A low Reynolds number ($Re<<1$) indicates that viscous forces are very strong compared to inertial forces. In such cases, inertial forces are sometimes neglected; this flow regime is called Stokes or creeping flow.

In contrast, high Reynolds numbers ($Re>>1$) indicate that the inertial effects have more effect on the velocity field than the viscous (friction) effects. In high Reynolds number flows, the flow is often modeled as an inviscid flow, an approximation in which viscosity is completely neglected. Eliminating viscosity allows the Navier-Stokes equations to be simplified into the Euler equations. The integration of the Euler equations along a streamline in an inviscid flow yields Bernoulli's equation. When, in addition to being inviscid, the flow is irrotational everywhere, Bernoulli's equation can completely describe the flow everywhere. Such flows are called potential flows, because the velocity field may be expressed as the gradient of a potential energy expression.

This idea can work fairly well when the Reynolds number is high. However, problems such as those involving solid boundaries may require that the viscosity be included. Viscosity cannot be neglected near solid boundaries because the no-slip condition generates a thin region of large strain rate, the boundary layer, in which viscosity effects dominate and which thus generates vorticity. Therefore, to calculate net forces on bodies (such as wings), viscous flow equations must be used: inviscid flow theory fails to predict drag forces, a limitation known as the d'Alembert's paradox.

A commonly used model, especially in computational fluid dynamics, is to use two flow models: the Euler equations away from the body, and boundary layer equations in a region close to the body. The two solutions can then be matched with each other, using the method of matched asymptotic expansions.

Steady vs Unsteady Flow

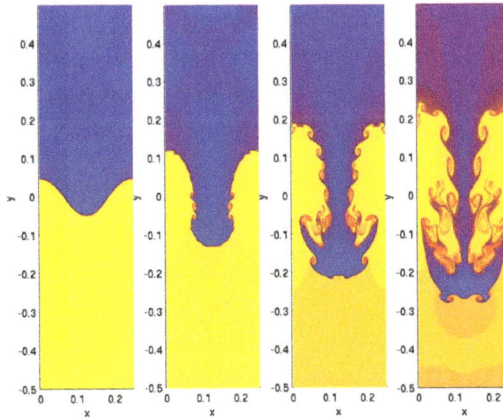

Hydrodynamics simulation of the Rayleigh–Taylor instability

A flow that is not a function of time is called steady flow. Steady-state flow refers to the condition where the fluid properties at a point in the system do not change over time. Time dependent flow is known as unsteady (also called transient). Whether a particular flow is steady or unsteady, can depend on the chosen frame of reference. For instance, laminar flow over a sphere is steady in the frame of reference that is stationary with respect to the sphere. In a frame of reference that is stationary with respect to a background flow, the flow is unsteady.

Turbulent flows are unsteady by definition. A turbulent flow can, however, be statistically stationary. According to Pope:

The random field $U(x,t)$ is statistically stationary if all statistics are invariant under a shift in time.

This roughly means that all statistical properties are constant in time. Often, the mean field is the object of interest, and this is constant too in a statistically stationary flow.

Steady flows are often more tractable than otherwise similar unsteady flows. The governing equations of a steady problem have one dimension fewer (time) than the governing equations of the same problem without taking advantage of the steadiness of the flow field.

Laminar vs Turbulent Flow

Turbulence is flow characterized by recirculation, eddies, and apparent randomness.

Flow in which turbulence is not exhibited is called laminar. It should be noted, however, that the presence of eddies or recirculation alone does not necessarily indicate turbulent flow—these phenomena may be present in laminar flow as well. Mathematically, turbulent flow is often represented via a Reynolds decomposition, in which the flow is broken down into the sum of an average component and a perturbation component.

It is believed that turbulent flows can be described well through the use of the Navier–Stokes equations. Direct numerical simulation (DNS), based on the Navier–Stokes equations, makes it possible to simulate turbulent flows at moderate Reynolds numbers. Restrictions depend on the power of the computer used and the efficiency of the solution algorithm. The results of DNS have been found to agree well with experimental data for some flows.

Most flows of interest have Reynolds numbers much too high for DNS to be a viable option, given the state of computational power for the next few decades. Any flight vehicle large enough to carry a human (L > 3 m), moving faster than 720 km/h (20 m/s) is well beyond the limit of DNS simulation (Re = 4 million). Transport aircraft wings (such as on an Airbus A300 or Boeing 747) have Reynolds numbers of 40 million (based on the wing chord dimension). Solving these real-life flow problems requires turbulence models for the foreseeable future. Reynolds-averaged Navier–Stokes equations (RANS) combined with turbulence modelling provides a model of the effects of the turbulent flow. Such a modelling mainly provides the additional momentum transfer by the Reynolds stresses, although the turbulence also enhances the heat and mass transfer. Another promising methodology is large eddy simulation (LES), especially in the guise of detached eddy simulation (DES)—which is a combination of RANS turbulence modelling and large eddy simulation.

Subsonic vs Transonic, Supersonic and Hypersonic Flows

While many flows (e.g. flow of water through a pipe) occur at low Mach numbers, many flows of practical interest in aerodynamics or in turbomachines occur at high fractions of M=1 (transonic flows) or in excess of it (supersonic or even hypersonic flows). New phenomena occur at these regimes such as instabilities in transonic flow, shock waves for supersonic flow, or non-equilibrium chemical behaviour due to ionization in hypersonic flows. In practice, each of those flow regimes is treated separately.

Reactive vs Non-reactive Flows

Reactive flows are flows that are chemically reactive, which finds its applications in many areas such as combustion(IC engine), propulsion devices(Rockets, jet engines etc.), detonations, fire and safety hazards, astrophysics etc. In addition to conservation of mass, momentum and energy, conservation of individual species (for example, mass fraction of methane in methane combustion) need to be derived, where the production/depletion rate of any species are obtained by simultaneously solving the equations of chemical kinetics.

Magnetohydrodynamics

Magnetohydrodynamics is the multi-disciplinary study of the flow of electrically conducting fluids in electromagnetic fields. Examples of such fluids include plasmas, liquid metals, and salt water. The fluid flow equations are solved simultaneously with Maxwell's equations of electromagnetism.

Other Approximations

There are a large number of other possible approximations to fluid dynamic problems. Some of the more commonly used are listed below.

- The *Boussinesq approximation* neglects variations in density except to calculate buoyancy forces. It is often used in free convection problems where density changes are small.

- *Lubrication theory* and *Hele–Shaw flow* exploits the large aspect ratio of the domain to show that certain terms in the equations are small and so can be neglected.

- *Slender-body theory* is a methodology used in Stokes flow problems to estimate the force on, or flow field around, a long slender object in a viscous fluid.

- The *shallow-water equations* can be used to describe a layer of relatively inviscid fluid with a free surface, in which surface gradients are small.

- *Darcy's law* is used for flow in porous media, and works with variables averaged over several pore-widths.

- In rotating systems, the *quasi-geostrophic equations* assume an almost perfect balance between pressure gradients and the Coriolis force. It is useful in the study of atmospheric dynamics.

Terminology in Fluid Dynamics

The concept of pressure is central to the study of both fluid statics and fluid dynamics. A pressure can be identified for every point in a body of fluid, regardless of whether the fluid is in motion or not. Pressure can be measured using an aneroid, Bourdon tube, mercury column, or various other methods.

Some of the terminology that is necessary in the study of fluid dynamics is not found in other similar areas of study. In particular, some of the terminology used in fluid dynamics is not used in fluid statics.

Terminology in Incompressible Fluid Dynamics

The concepts of total pressure and dynamic pressure arise from Bernoulli's equation and are significant in the study of all fluid flows. (These two pressures are not pressures in the usual sense—they cannot be measured using an aneroid, Bourdon tube or mercury column.) To avoid potential ambiguity when referring to pressure in fluid dynamics, many authors use the term static pressure to distinguish it from total pressure and dynamic pressure. Static pressure is identical to pressure and can be identified for every point in a fluid flow field.

In *Aerodynamics*, L.J. Clancy writes: *To distinguish it from the total and dynamic pressures, the actual pressure of the fluid, which is associated not with its motion but with its state, is often referred to as the static pressure, but where the term pressure alone is used it refers to this static pressure.*

A point in a fluid flow where the flow has come to rest (i.e. speed is equal to zero adjacent to some solid body immersed in the fluid flow) is of special significance. It is of such importance that it is given a special name—a stagnation point. The static pressure at the stagnation point is of special significance and is given its own name—stagnation pressure. In incompressible flows, the stagnation pressure at a stagnation point is equal to the total pressure throughout the flow field.

Terminology in Compressible Fluid Dynamics

In a compressible fluid, it is convenient to define the total conditions (also called stagnation conditions) for all thermodynamic state properties (e.g. total temperature, total enthalpy, total speed of sound). These total flow conditions are a function of the fluid velocity and have different values in frames of reference with different motion.

To avoid potential ambiguity when referring to the properties of the fluid associated with the state of the fluid rather than its motion, the prefix "static" is commonly used (e.g. static temperature, static enthalpy). Where there is no prefix, the fluid property is the static condition (i.e. "density" and "static density" mean the same thing). The static conditions are independent of the frame of reference.

Because the total flow conditions are defined by isentropically bringing the fluid to rest, there is no need to distinguish between total entropy and static entropy as they are always equal by definition. As such, entropy is most commonly referred to as simply "entropy".

Flow Measurement

Flow measurement is the quantification of bulk fluid movement. Flow can be measured in a variety of ways. Positive-displacement flow meters accumulate a fixed volume of fluid and then count the number of times the volume is filled to measure flow. Other flow measurement methods rely on forces produced by the flowing stream as it overcomes a known constriction, to indirectly calculate flow. Flow may be measured by measuring the velocity of fluid over a known area.

Units of Measurement

Both gas and liquid flow can be measured in volumetric or mass flow rates, such as liters per second or kilograms per second, respectively. These measurements are related by the material's density. The density of a liquid is almost independent of conditions. This is not the case for gases, the densities of which depend greatly upon pressure, temperature and to a lesser extent, composition.

When gases or liquids are transferred for their energy content, as in the sale of natural gas, the flow rate may also be expressed in terms of energy flow, such as GJ/hour or BTU/day. The energy flow rate is the volumetric flow rate multiplied by the energy content per unit volume or mass flow rate multiplied by the energy content per unit mass. Energy flow rate is usually derived from mass or volumetric flow rate by the use of a flow computer.

In engineering contexts, the volumetric flow rate is usually given the symbol Q, and the mass flow rate, the symbol $\dot{m}$.

For a fluid having density ρ, mass and volumetric flow rates may be related by $\dot{m} = \rho * Q$.

Gas

Gases are compressible and change volume when placed under pressure, are heated or are cooled. A volume of gas under one set of pressure and temperature conditions is not equivalent to the same gas under different conditions. References will be made to "actual" flow rate through a meter and "standard" or "base" flow rate through a meter with units such as *acm/h* (actual cubic meters per hour), *sm³/sec* (standard cubic meters per second), *kscm/h* (thousand standard cubic meters per hour), *LFM* (linear feet per minute), or *MMSCFD* (million standard cubic feet per day).

Gas *mass* flow rate can be directly measured, independent of pressure and temperature effects, with thermal mass flow meters, Coriolis mass flow meters, or mass flow controllers.

Liquid

For liquids, various units are used depending upon the application and industry, but might include gallons (U.S. or imperial) per minute, liters per second, bushels per minute or, when describing river flows, cumecs (cubic metres per second) or acre-feet per day. In oceanography a common unit to measure volume transport (volume of water transported by a current for example) is a sverdrup (Sv) equivalent to 10^6 m³/s.

Mechanical Flow Meters

A positive displacement meter may be compared to a bucket and a stopwatch. The stopwatch is started when the flow starts, and stopped when the bucket reaches its limit. The volume divided by the time gives the flow rate. For continuous measurements, we need a system of continually filling and emptying buckets to divide the flow without letting it out of the pipe. These continuously forming and collapsing volumetric displacements may take the form of pistons reciprocating in cylinders, gear teeth mating against the internal wall of a meter or through a progressive cavity created by rotating oval gears or a helical screw.

Piston Meter/Rotary Piston

Because they are used for domestic water measurement, piston meters, also known as rotary piston or semi-positive displacement meters, are the most common flow measurement devices in the UK and are used for almost all meter sizes up to and including 40 mm (1 $\frac{1}{2}$ in). The piston meter operates on the principle of a piston rotating within a chamber of known volume. For each rotation, an amount of water passes through the piston chamber. Through a gear mechanism and, sometimes, a magnetic drive, a needle dial and odometer type display are advanced.

Oval Gear Meter

A positive displacement flowmeter of the oval gear type. Fluid forces the meshed gears to rotate; each rotation corresponds to a fixed volume of fluid. Counting the revolutions totalizes volume, and the rate is proportional to flow

An oval gear meter is a positive displacement meter that uses two or more oblong gears configured to rotate at right angles to one another, forming a T shape. Such a meter has two sides, which can be called A and B. No fluid passes through the center of the meter, where the teeth of the two gears always mesh. On one side of the meter (A), the teeth of the gears close off the fluid flow because the elongated gear on side A is protruding into the measurement chamber, while on the other side of the meter (B), a cavity holds a fixed volume of fluid in a measurement chamber. As the fluid pushes the gears, it rotates them, allowing the fluid in the measurement chamber on side B to be released into the outlet port. Meanwhile, fluid entering the inlet port will be driven into the measurement chamber of side A, which is now open. The teeth on side B will now close off the fluid from entering side B. This cycle continues as the gears rotate and fluid is metered through alternating measurement chambers. Permanent magnets in the rotating gears can transmit a signal to an electric reed switch or current transducer for flow measurement. Though claims for high performance are made, they are generally not as precise as the sliding vane design.

Gear Meter

Gear meters differ from Oval Gear meters in that the measurement chambers are made up of the gaps between the teeth of the gears. These openings divide up the fluid stream and as the gears rotate away from the inlet port, the meter's inner wall closes off the chamber to hold the fixed amount of fluid. The outlet port is located in the area where the gears are coming back together. The fluid is forced out of the meter as the gear teeth mesh and reduce the available pockets to nearly zero volume.

Helical Gear

Helical gear flow meters get their name from the shape of their gears or rotors. These rotors resemble the shape of a helix, which is a spiral-shaped structure. As the fluid flows through the meter, it enters the compartments in the rotors, causing the rotors to rotate. The length of the rotor is sufficient that the inlet and outlet are always separated from each other thus blocking a free flow of liquid. The mating helical rotors create a progressive cavity which opens to admit fluid, seals itself off and then opens up to the downstream side to release the fluid. This happens in a continuous fashion and the flowrate is calculated from the speed of rotation.

Nutating Disk Meter

This is the most commonly used measurement system for measuring water supply in houses. The fluid, most commonly water, enters in one side of the meter and strikes the nutating disk, which is eccentrically mounted. The disk must then "wobble" or nutate about the vertical axis, since the bottom and the top of the disk remain in contact with the mounting chamber. A partition separates the inlet and outlet chambers. As the disk nutates, it gives direct indication of the volume of the liquid that has passed through the meter as volumetric flow is indicated by a gearing and register arrangement, which is connected to the disk. It is reliable for flow measurements within 1 percent.

Variable Area Meter

The variable area (VA) meter, also commonly called a rotameter, consists of a tapered tube, typically made of glass, with a float inside that is pushed up by fluid flow and pulled down by gravity. As flow rate increases, greater viscous and pressure forces on the float cause it to rise until it becomes stationary at a location in the tube that is wide enough for the forces to balance. Floats are made in many different shapes, with spheres and spherical ellipses being the most common. Some are designed to spin visibly in the fluid stream to aid the user in determining whether the float is stuck or not. Rotameters are available for a wide range of liquids but are most commonly used with water or air. They can be made to reliably measure flow down to 1% accuracy.

Turbine Flow Meter

The turbine flow meter (better described as an axial turbine) translates the mechanical action of the turbine rotating in the liquid flow around an axis into a user-readable rate of flow (gpm, lpm, etc.). The turbine tends to have all the flow traveling around it.

The turbine wheel is set in the path of a fluid stream. The flowing fluid impinges on the turbine blades, imparting a force to the blade surface and setting the rotor in motion. When a steady rotation speed has been reached, the speed is proportional to fluid velocity.

Turbine flow meters are used for the measurement of natural gas and liquid flow. Turbine meters are less accurate than displacement and jet meters at low flow rates, but the measuring element does not occupy or severely restrict the entire path of flow. The flow direction is generally straight through the meter, allowing for higher flow rates and less pressure loss than displacement-type meters. They are the meter of choice for large commercial users, fire protection, and as master meters for the water distribution system. Strainers are generally required to be installed in front of the meter to protect the measuring element from gravel or other debris that could enter the water distribution system. Turbine meters are generally available for 4 to 30 cm (1 ½–12 in) or higher pipe sizes. Turbine meter bodies are commonly made of bronze, cast Iron, or ductile iron. Internal turbine elements can be plastic or non-corrosive metal alloys. They are accurate in normal working conditions but are greatly affected by the flow profile and fluid conditions.

Fire meters are a specialized type of turbine meter with approvals for the high flow rates required in fire protection systems. They are often approved by Underwriters Laboratories (UL) or Factory Mutual (FM) or similar authorities for use in fire protection. Portable turbine meters may be temporarily installed to measure water used from a fire hydrant. The meters are normally made of aluminum to be lightweight, and are usually 7.5 cm (3 in) capacity. Water utilities often require them for measurement of water used in construction, pool filling, or where a permanent meter is not yet installed.

Woltman Meter

The Woltman meter (invented by Reinhard Woltman in the 19th century) comprises a rotor with helical blades inserted axially in the flow, much like a ducted fan; it can be considered a type of turbine flow meter. They are commonly referred to as helix meters, and are popular at larger sizes.

Single Jet Meter

A single jet meter consists of a simple impeller with radial vanes, impinged upon by a single jet. They are increasing in popularity in the UK at larger sizes and are commonplace in the EU.

Paddle Wheel Meter

This is similar to the single jet meter, except that the impeller is small with respect to the width of the pipe, and projects only partially into the flow, like the paddle wheel on a Mississippi riverboat.

Multiple Jet Meter

A multiple jet or multijet meter is a velocity type meter which has an impeller which rotates horizontally on a vertical shaft. The impeller element is in a housing in which multiple inlet ports direct the fluid flow at the impeller causing it to rotate in a specific direction in proportion to the flow velocity. This meter works mechanically much like a single jet meter except that the ports direct the flow at the impeller equally from several points around the circumference of the element, not just one point; this minimizes uneven wear on the impeller and its shaft. Thus these types of meters are recommended to be installed horizontally with its roller index pointing skywards.

Pelton Wheel

The Pelton wheel turbine (better described as a radial turbine) translates the mechanical action of the Pelton wheel rotating in the liquid flow around an axis into a user-readable rate of flow (gpm, lpm, etc.). The Pelton wheel tends to have all the flow traveling around it with the inlet flow focused on the blades by a jet. The original Pelton wheels were used for the generation of power and consisted of a radial flow turbine with "reaction cups" which not only move with the force of the water on the face but return the flow in opposite direction using this change of fluid direction to further increase the efficiency of the turbine.

Current Meter

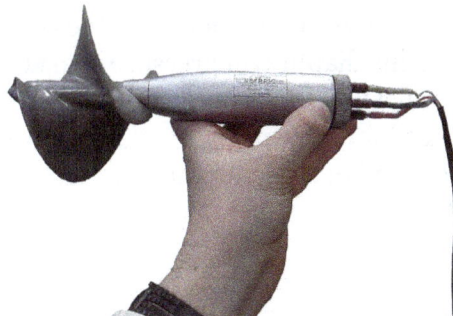

A propeller-type current meter as used for hydroelectric turbine testing

Flow through a large penstock such as used at a hydroelectric power plant can be measured by averaging the flow velocity over the entire area. Propeller-type current meters (similar to the purely mechanical Ekman current meter, but now with electronic data acquisition) can be traversed over the area of the penstock and velocities averaged to calculate total flow. This may be on the order of hundreds of cubic meters per second. The flow must be kept steady during the traverse of the current meters. Methods for testing hydroelectric turbines are given in IEC standard 41. Such flow measurements are often commercially important when testing the efficiency of large turbines.

Pressure-based Meters

There are several types of flow meter that rely on Bernoulli's principle, either by measuring the differential pressure within a constriction, or by measuring static and stagnation pressures to derive the dynamic pressure.

Venturi Meter

A Venturi meter constricts the flow in some fashion, and pressure sensors measure the differential pressure before and within the constriction. This method is widely used to measure flow rate in the transmission of gas through pipelines, and has been used since Roman Empire times. The coefficient of discharge of Venturi meter ranges from 0.93 to 0.97. The first large-scale Venturi meters to measure liquid flows were developed by Clemens Herschel who used them to measure small and large flows of water and wastewater beginning at the end of the 19th century.

Orifice Plate

An orifice plate is a plate with a hole through it, placed in the flow; it constricts the flow, and measuring the pressure differential across the constriction gives the flow rate. It is basically a crude form of Venturi meter, but with higher energy losses. There are three type of orifice: concentric, eccentric, and segmental.

Dall Tube

The Dall tube is a shortened version of a Venturi meter, with a lower pressure drop than an orifice plate. As with these flow meters the flow rate in a Dall tube is determined by measuring the pressure drop caused by restriction in the conduit. The pressure differential is typically measured using diaphragm pressure transducers with digital readout. Since these meters have significantly lower permanent pressure losses than orifice meters, Dall tubes are widely used for measuring the flow rate of large pipeworks. Differential pressure produced by a Dall tube higher than Venturi tube and nozzle, all of them having same throat diameters.

Pitot-tube

A Pitot-tube is a pressure measuring instrument used to measure fluid flow velocity by determining the stagnation pressure and static pressure. Bernoulli's equation used to calculate the dynamic pressure and hence fluid velocity.

Multi-hole Pressure Probe

Multi-hole pressure probes (also called impact probes) extend the theory of Pitot tube to more than one dimension. A typical impact probe consists of three or more holes (depending on the type of probe) on the measuring tip arranged in a specific pattern. More holes allow the instrument to measure the direction of the flow velocity in addition to its magnitude (after appropriate calibration). Three holes arranged in a line allow the pressure probes to measure the velocity vector in two dimensions. Introduction of more holes, e.g. five holes arranged in a "plus" formation, allow measurement of the three-dimensional velocity vector.

Cone Meters

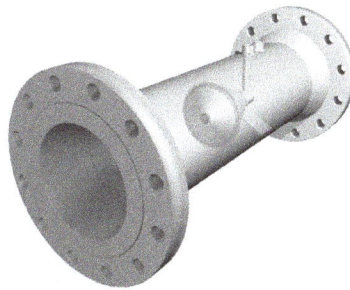

8inch (200mm) V-cone flowmeter shown with ANSI 300# raised face weld neck flanges

Cone meters are a newer differential pressure metering device first launched in 1985 by McCrometer in Hemet, CA. While working with the same basic principles as Venturi and orifice type DP meters, cone meters don't require the same upstream and downstream piping. The cone acts as a conditioning device as well as a differential pressure producer. Upstream requirements are between 0–5 diameters compared to up to 44 diameters for an orifice plate or 22 diameters for a Venturi. Because cone meters are generally of welded construction, it is recommended they are always calibrated prior to service. Inevitably heat effects of welding cause distortions and other effects that prevent tabular data on discharge coefficients with respect to line size, beta ratio and operating Reynolds numbers from being collected and published. Calibrated cone meters have an uncertainty up to +/-0.5%. Un-calibrated cone meters have an uncertainty of +/-5.0%.

Linear Resistance Meters

Linear resistance meters, also called laminar flow meters, measure very low flows at which the measured differential pressure is linearly proportional to the flow and to the fluid viscosity. Such flow is called viscous drag flow or laminar flow, as opposed to the turbulent flow measured by orifice plates, Venturis and other meters mentioned in this section, and is characterized by Reynolds numbers below 2000. The primary flow element may consist of a single long capillary tube, a bundle of such tubes, or a long porous plug; such low flows create small pressure differentials but longer flow elements create higher, more easily measured differentials. These flow meters are particularly sensitive to temperature changes affecting the fluid viscosity and the diameter of the flow element, as can be seen in the governing Hagen-Poiseuille equation.

Optical Flow Meters

Optical flow meters use light to determine flow rate. Small particles which accompany natural and industrial gases pass through two laser beams focused a short distance apart in the flow path. in a pipe by illuminating optics. Laser light is scattered when a particle crosses the first beam. The detecting optics collects scattered light on a photo-detector, which then generates a pulse signal. As the same particle crosses the second beam, the detecting optics collect scattered light on a second photodetector, which converts the incoming light into a second electrical pulse. By measuring the time interval between these pulses, the gas velocity is calculated as $V = D / t$ where D is the distance between the laser beams and t is the time interval.

Laser-based optical flow meters measure the actual speed of particles, a property which is not dependent on thermal conductivity of gases, variations in gas flow or composition of gases. The operating principle enables optical laser technology to deliver highly accurate flow data, even in challenging environments which may include high temperature, low flow rates, high pressure, high humidity, pipe vibration and acoustic noise.

Optical flow meters are very stable with no moving parts and deliver a highly repeatable measurement over the life of the product. Because distance between the two laser sheets does not change, optical flow meters do not require periodic calibration after their initial commissioning. Optical flow meters require only one installation point, instead of the two installation points typically required by other types of meters. A single installation point is simpler, requires less maintenance and is less prone to errors.

Commercially available optical flow meters are capable of measuring flow from 0.1 m/s to faster than 100 m/s (1000:1 turn down ratio) and have been demonstrated to be effective for the measurement of flare gases from oil wells and refineries, a contributor to atmospheric pollution.

Open-channel Flow Measurement

Open channel flow describes cases where flowing liquid has a top surface open to the air; the cross-section of the flow is only determined by the shape of the channel on the lower side, and is variable depending on the depth of liquid in the channel. Techniques appropriate for a fixed cross-section of flow in a pipe are not useful in open channels.

Level to Flow

The level of the water is measured at a designated point behind weir or in flume a hydraulic structure using various secondary devices (bubblers, ultrasonic, float, and differential pressure are common methods). This depth is converted to a flow rate according to a theoretical formula of the form $Q = KH^X$ where Q is the flow rate, K is a constant, H is the water level, and X is an exponent which varies with the device used; or it is converted according to empirically derived level/flow data points (a "flow curve"). The flow rate can then be integrated over time into volumetric flow. Level to flow devices are commonly used to measure the flow of surface waters (springs, stream, and rivers), industrial discharges, and sewage. Of these, weirs are used on flow streams with low solids (typically surface waters), while flumes are used on flows containing low or high solids contents.

Area/Velocity

The cross-sectional area of the flow is calculated from a depth measurement and the average velocity of the flow is measured directly (Doppler and propeller methods are common). Velocity times the cross-sectional area yields a flow rate which can be integrated into volumetric flow. There are two types of area velocity flow meter: (1) wetted; and (2) non-contact. Wetted area velocity sensors have to be typically mounted on the bottom of a channel or river and use Doppler to measure the velocity of the entrained particles. With depth and a programmed cross-section this can then provide discharge flow measurement. Non-contact devices that use laser or radar are mounted above the channel and measure the velocity from above and then use ultrasound to measure the depth of the water from above. Radar devices can only measure surface velocites, whereas laser-based devices can measure velocities sub-surface.

Dye Testing

A known amount of dye (or salt) per unit time is added to a flow stream. After complete mixing, the concentration is measured. The dilution rate equals the flow rates.

Acoustic Doppler Velocimetry

Acoustic Doppler velocimetry (ADV) is designed to record instantaneous velocity components at a single point with a relatively high frequency. Measurements are performed by measuring the velocity of particles in a remote sampling volume based upon the Doppler shift effect.

Thermal Mass Flow Meters

Thermal mass flow meters generally use combinations of heated elements and temperature sensors to measure the difference between static and flowing heat transfer to a fluid and infer its flow with a knowledge of the fluid's specific heat and density. The fluid temperature is also measured and compensated for. If the density and specific heat characteristics of the fluid are constant, the meter can provide a direct mass flow readout, and does not need any additional pressure temperature compensation over their specified range.

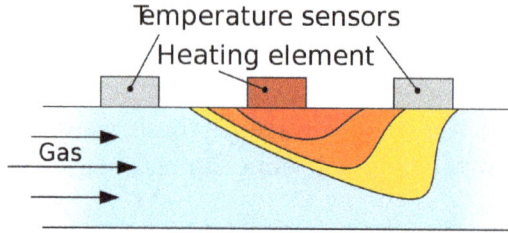

Temperature at the sensors varies depending upon the mass flow

Technological progress has allowed the manufacture of thermal mass flow meters on a microscopic scale as MEMS sensors; these flow devices can be used to measure flow rates in the range of nanolitres or microlitres per minute.

Thermal mass flow meter (also called thermal dispersion or thermal displacement flowmeter) technology is used for compressed air, nitrogen, helium, argon, oxygen, and natural gas. In fact, most gases can be measured as long as they are fairly clean and non-corrosive. For more aggressive gases, the meter may be made out of special alloys (e.g. Hastelloy), and pre-drying the gas also helps to minimize corrosion.

Today, thermal mass flow meters are used to measure the flow of gases in a growing range of applications, such as chemical reactions or thermal transfer applications that are difficult for other flow metering technologies. This is because thermal mass flow meters monitor variations in one or more of the thermal characteristics (temperature, thermal conductivity, and/or specific heat) of gaseous media to define the mass flow rate.

The MAF sensor

In many late model automobiles, a mass airflow sensor (MAF sensor) is used to accurately determine the mass flowrate of intake air used in the internal combustion engine. Many such mass flow sensors use a heated element and a downstream temperature sensor to indicate the air flowrate. Other sensors use a spring-loaded vane. In either case, the vehicle's electronic control unit interprets the sensor signals as a real time indication of an engine's fuel requirement.

Vortex Flow Meters

Another method of flow measurement involves placing a bluff body (called a shedder bar) in the path of the fluid. As the fluid passes this bar, disturbances in the flow called vortices are created. The vortices trail behind the cylinder, alternatively from each side of the bluff body. This vortex trail is called the Von Kármán vortex street after von Kármán's 1912 mathematical description of the phenomenon. The frequency at which these vortices alternate sides is essentially proportional to the flow rate of the fluid. Inside, atop, or downstream of the shedder bar is a sensor for measuring the frequency of the vortex shedding. This sensor is often a piezoelectric crystal, which produces a small, but measurable, voltage pulse every time a vortex is created. Since the frequency of such a voltage pulse is also proportional to the fluid velocity, a volumetric flow rate is calculated using the cross sectional area of the flow meter. The frequency is measured and the flow rate is calculated by the flowmeter electronics using the equation $f = SV / L$ where f is the frequency of the vortices, L the characteristic length of the bluff body, V is the velocity of the flow over the bluff body, and S is the Strouhal number, which is essentially a constant for a given body shape within its operating limits.

Sonar Flow Measurement

Sonar flow meter on gas line

Sonar flow meters are non-intrusive clamp on devices that measure flow in pipes conveying slurries, corrosive fluids, multiphase fluids and flows where insertion type flow meters are not desired. Sonar flow meters have been widely adopted in mining, metals processing, and upstream oil and gas industries where traditional technologies have certain limitations due to their tolerance to various flow regimes and turn down ratios.

Sonar flow meters have the capacity of measuring the velocity of liquids or gases non intrusively within the pipe and then leverage this velocity measurement into a flow rate by using the cross sectional area of the pipe and the line pressure and temperature. The principle behind this flow measurement is the use of underwater acoustics.

In underwater acoustics, to locate an object underwater, sonar uses two knowns:

- The speed of sound propagation through the array (i.e. the sound speed of sea water)

- The spacing between the sensors in the sensor array

and then calculates the unknown:

- The location (or angle) of the object.

Likewise, sonar flow measurement uses the same techniques and algorithms employed in underwater acoustics, but applies them to flow measurement of oil and gas wells and flow lines.

To measure flow velocity, sonar flow meters use two knowns:

- The location (or angle) of the object, which is 0 degrees since the flow is moving along the pipe, which is aligned with the sensor array

- The spacing between the sensors in the sensor array

and then calculates the unknown:

- The speed of propagation through the array (i.e. the flow velocity of the medium in the pipe).

Electromagnetic, Ultrasonic and Coriolis Flow Meters

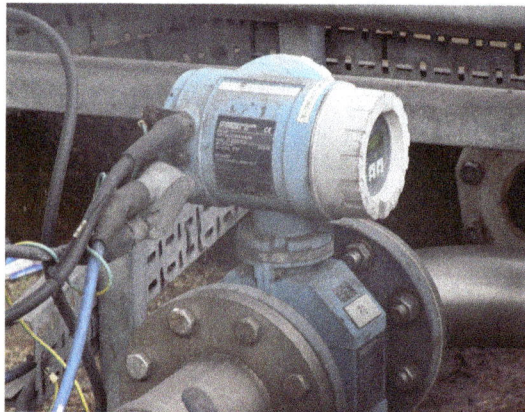

A magnetic flow meter at the Tetley's Brewery in Leeds, West Yorkshire

Modern innovations in the measurement of flow rate incorporate electronic devices that can correct for varying pressure and temperature (i.e. density) conditions, non-linearities, and for the characteristics of the fluid.

Magnetic Flow Meters

Magnetic flow meters, often called "mag meter" or "electromag", use a magnetic field applied to the metering tube, which results in a potential difference proportional to the flow velocity perpendicular to the flux lines. The potential difference is sensed by electrodes aligned perpendicular to the flow and the applied magnetic field. The physical principle at work is Faraday's law of electromagnetic induction. The magnetic flow meter requires a conducting fluid and a nonconducting pipe liner. The electrodes must not corrode in contact with the process fluid; some magnetic flowmeters have auxiliary transducers installed to clean the electrodes in place. The applied magnetic field is pulsed, which allows the flowmeter to cancel out the effect of stray voltage in the piping system.

Non-contact Electromagnetic Flow Meters

A Lorentz force velocimetry system is called Lorentz force flowmeter (LFF). A LFF measures the integrated or bulk Lorentz force resulting from the interaction between a liquid metal in motion and an applied magnetic field. In this case the characteristic length of the magnetic field is of the same order of magnitude as the dimensions of the channel. It must be addressed that in the case where localized magnetic fields are used, it is possible to perform local velocity measurements and thus the term Lorentz force velocimeter is used.

Ultrasonic Flow Meters (Doppler, Transit Time)

There are two main types of ultrasonic flow meters: Doppler and transit time. While they both utilize ultrasound to make measurements and can be non-invasive (measure flow from outside the tube, pipe or vessel), they measure flow by very different methods.

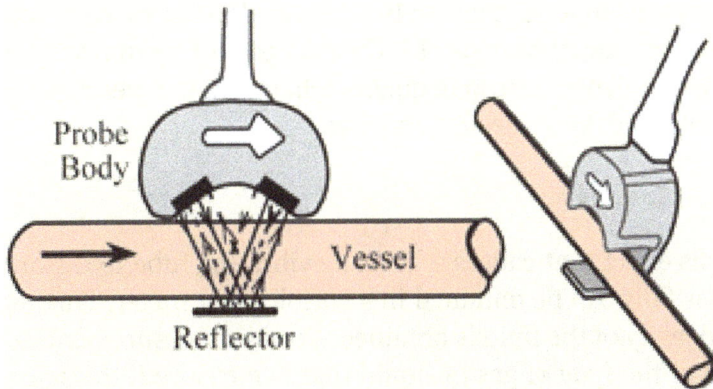

Schematic view of a flow sensor.

Ultrasonic transit time flow meters measure the difference of the transit time of ultrasonic pulses propagating in and against the direction of flow. This time difference is a measure for the average velocity of the fluid along the path of the ultrasonic beam. By using the absolute transit times both the averaged fluid velocity and the speed of sound can be calculated. Using the two transit times t_{up} and t_{down} and the distance between receiving and transmitting transducers L and the inclination angle α one can write the equations:

$$v = \frac{L}{2\sin(\alpha)} \frac{t_{up} - t_{down}}{t_{up}\, t_{down}} \quad \text{and} \quad c = \frac{L}{2} \frac{t_{up} + t_{down}}{t_{up}\, t_{down}}$$

where v is the average velocity of the fluid along the sound path and c is the speed of sound.

With wide-beam illumination transit time ultrasound can also be used to measure volume flow independent of the cross-sectional area of the vessel or tube.

Ultrasonic Doppler flow meters measure the Doppler shift resulting from reflecting an ultrasonic beam off the particulates in flowing fluid. The frequency of the transmitted beam is affected by the movement of the particles; this frequency shift can be used to calculate the fluid velocity. For the Doppler principle to work there must be a high enough density of sonically reflective materials such as solid particles or air bubbles suspended in the fluid. This is in direct contrast to an ultrasonic transit time flow meter, where bubbles and solid particles reduce the accuracy of the measurement. Due to the dependency on these particles there are limited applications for Doppler flow meters. This technology is also known as acoustic Doppler velocimetry.

One advantage of ultrasonic flow meters is that they can effectively measure the flow rates for a wide variety of fluids, as long as the speed of sound through that fluid is known. For example, ultrasonic flow meters are used for the measurement of such diverse fluids a liquid natural gas (LNG) and blood. One can also calculate the expected speed of sound for a given fluid; this can be compared to the speed of sound empirically measured by an ultrasonic flow meter for the purposes of monitoring the quality of the flow meter's measurements. A drop in quality (change in the measured speed of sound) is an indication that the meter needs servicing.

Coriolis Flow Meters

Using the Coriolis effect that causes a laterally vibrating tube to distort, a direct measurement of mass flow can be obtained in a coriolis flow meter. Furthermore, a direct measure of the density of the fluid is obtained. Coriolis measurement can be very accurate irrespective of the type of gas or liquid that is measured; the same measurement tube can be used for hydrogen gas and bitumen without recalibration.

Coriolis flow meters can be used for the measurement of natural gas flow.

Laser Doppler Flow Measurement

A beam of laser light impinging on a moving particle will be partially scattered with a change in wavelength proportional to the particle's speed (the Doppler effect). A laser Doppler velocimeter (LDV), also called a laser Doppler anemometer (LDA), focuses a laser beam into a small volume in a flowing fluid containing small particles (naturally occurring or induced). The particles scatter the light with a Doppler shift. Analysis of this shifted wavelength can be used to directly, and with great precision, determine the speed of the particle and thus a close approximation of the fluid velocity.

A number of different techniques and device configurations are available for determining the Doppler shift. All use a photodetector (typically an avalanche photodiode) to convert the light into an electrical waveform for analysis. In most devices, the original laser light is divided into two beams. In one general LDV class, the two beams are made to intersect at their focal points where they interfere and generate a set of straight fringes. The sensor is then aligned to the flow such that the fringes are perpendicular to the flow direction. As particles pass through the fringes, the Doppler-shifted light is collected into the photodetector. In another general LDV class, one beam is used as a reference and the other is Doppler-scattered. Both beams are then collected onto the photodetector where optical heterodyne detection is used to extract the Doppler signal.

Calibration

Even though ideally the flowmeter should be unaffected by its environment, in practice this is unlikely to be the case. Often measurement errors originate from incorrect installation or other environment dependent factors. In situ methods are used when flow meter is calibrated in the correct flow conditions.

Transit Time Method

For pipe flows a so-called transit time method is applied where a radiotracer is injected as a pulse into the measured flow. The transit time is defined with the help of radiation detectors placed on the outside of the pipe. The volume flow is obtained by multiplying the measured average fluid flow velocity by the inner pipe cross section. This reference flow value is compared with the simultaneous flow value given by the flow measurement to be calibrated.

The procedure is standardised (ISO 2975/VII for liquids and BS 5857-2.4 for gases). The best accredited measurement uncertainty for liquids and gases is 0.5%.

- Transit time animation

Tracer Dilution Method

The radiotracer dilution method is used to calibrate open channel flow measurements. A solution with a known tracer concentration is injected at a constant known velocity into the channel flow. Downstream the tracer solution is thoroughly mixed over the flow cross section, a continuous sample is taken and its tracer concentration in relation to that of the injected solution is determined. The flow reference value is determined by using the tracer balance condition between the injected tracer flow and the diluting flow. The procedure is standardised (ISO 9555-1 and ISO 9555-2 for liquid flow in open channels). The best accredited measurement uncertainty is 1%.

- Tracer dilution animation

Steady Flow

In this lesson we will obtain the solution of the steady state flow problem in confined and unconfined aquifer.

Confined aquifer

In case of steady flow in confined aquifer, the flow equation becomes

Or, $\dfrac{1}{r}\dfrac{d}{dr}\left(r\dfrac{dh}{dr}\right) = 0$

Or, $\dfrac{d}{dr}\left(r\dfrac{dh}{dr}\right) = 0$

Integrating, $\displaystyle\int \dfrac{d}{dr}\left(r\dfrac{dh}{dr}\right) = \int 0$

Or, $r\dfrac{dh}{dr} = C_1$

Now, Darcy's law can be expressed as

$Q = 2\pi r T \dfrac{dh}{dr} \rightarrow \dfrac{Q}{2\pi T} = r\dfrac{dh}{dr} = C_1$

Therefore, the equation above can be written as

$r\dfrac{dh}{dr} = \dfrac{Q}{2\pi T}$

$\Rightarrow dh = \dfrac{Q}{2\pi T}\dfrac{dr}{r}$

Now integrating, we have

$\Rightarrow \displaystyle\int dh = \dfrac{Q}{2\pi T}\int \dfrac{dr}{r}$

$\Rightarrow h = \dfrac{Q}{2\pi T}\ln(r)C_2$

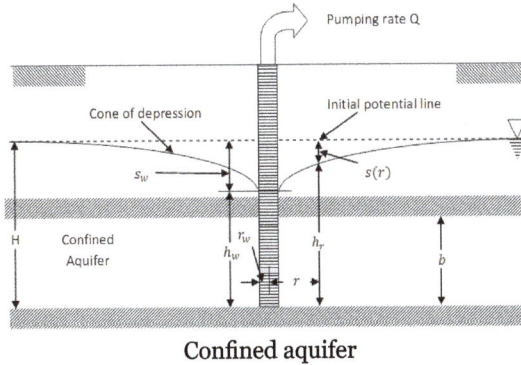

Confined aquifer

Now consider the figure above:

$$For\ r = r_w \quad \rightarrow \quad h = h_w$$
$$For\ r = r \quad \rightarrow \quad h = h_r$$

Putting it in equation above:

$$h_w = \frac{Q}{2\pi T}\ln\left(r_w\right) + C_2$$

And,

$$h_r = \frac{Q}{2\pi T}\ln\left(r\right) + C_2$$

From these two equations, we have

$$h_r - h_w = \frac{Q}{2\pi T}\ln\left(r\right) - \frac{Q}{2\pi T}\ln\left(r_w\right)$$

$$h_r - h_w = \frac{Q}{2\pi T}\ln\left(\frac{r}{r_w}\right)$$

Knowing hydraulic head at the well, the equation above can be used to calculate steady state hydraulic head for any values of r. This equation can also be used for estimation of aquifer transmissivity. For calculating aquifer transmissivity, the equation can be written as,

$$T = \frac{Q}{2\pi\left(h_r - h_w\right)}\ln\left(\frac{r}{r_w}\right)$$

Unconfined Aquifer

In case of steady flow in unconfined aquifer, the flow equation becomes

Or, $\dfrac{1}{r}\dfrac{d}{dr}\left(rh\dfrac{dh}{dr}\right) = 0$

Or, $\dfrac{d}{dr}\left(rh\dfrac{dh}{dr}\right) = 0$

Integrating, $\int \dfrac{d}{dr}\left(rh\dfrac{dh}{dr}\right) = \int 0$

Or, $rh\dfrac{dh}{dr} = C_1$

Now, Darcy's law can be expressed as

$Q = 2\pi r K h\dfrac{dh}{dr} \rightarrow \dfrac{Q}{2\pi K} = rh\dfrac{dh}{dr} = C_1$

Therefore, the equation can be written as

$rh\dfrac{dh}{dr} = \dfrac{Q}{2\pi T}$

$\Rightarrow hdh = \dfrac{Q}{2\pi K}\dfrac{dr}{r}$

Now integrating, we have

$\Rightarrow \int hdh = \dfrac{Q}{2\pi K}\int \dfrac{dr}{r}$

$\Rightarrow \dfrac{h^2}{2} = \dfrac{Q}{2\pi K}\ln(r) + C_2$

$\Rightarrow h^2 = \dfrac{Q}{\pi K}\ln(r) + C_2$

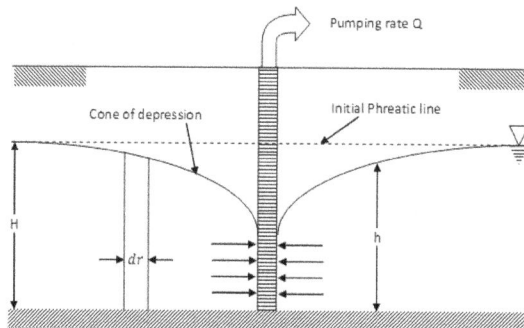

An unconfined aquifer with boundary condition

Consider the unconfined aquifer shown in above figure:

$$For\ r = r_w \rightarrow \quad h = h_w$$
$$For\ r = r \rightarrow \quad h = h_r$$

Putting it in the equation

$$\Rightarrow h_w^2 = \frac{Q}{\pi K} \ln(r_w) + C_2$$

And,

$$\Rightarrow h_r^2 = \frac{Q}{\pi K} \ln(r_r) + C_2$$

From these two equations, we have

$$h_r^2 - h_w^2 = \frac{Q}{\pi K} \ln(r_r) - \frac{Q}{\pi K} \ln(r_w)$$

$$\Rightarrow h_r^2 - h_w^2 = \frac{Q}{\pi K} \ln\left(\frac{r_r}{r_w}\right)$$

Knowing hydraulic head at the well, the equation above can be used to calculate steady hydraulic head for any values of r. This equation can also be used for estimation of aquifer conductivity. The equation can be written for calculating aquifer conductivity as,

$$K = \frac{Q}{\pi \left(h_r^2 - h_w^2\right)} \ln\left(\frac{r_r}{r_w}\right)$$

Unsteady Flow

We have already derived the flow equation for unsteady flow in confined aquifer. The equation can be written as,

$$\frac{\partial^2 h}{\partial r^2} + \frac{1}{r}\frac{\partial h}{\partial r} = \frac{S_S}{T}\frac{\partial h}{\partial t}$$

Theis (1935) obtained the solution of the equation. His solution was based on the analogy between groundwater flow and heat conduction. Considering the following boundary conditions,

at $t = 0$ $h = h_0$

at $t = \infty$ $h = h_0$

The solution of the equation for $t \geq o$ is

$$s(r,t) = \frac{Q}{4\pi T} W(u)$$

Where, $s(r,t)$ is the draw down at a radial distance r from, the well at time t,

$$u = \frac{r^2 S_s}{4Tt} \text{ and } W(u) = \int_u^\infty \frac{e^{-u}}{u} du$$

$W(u)$ is the exponential integration and is known as well function. The well function $W(u)$ can be approximated as

$$W(u) = -0.5772 - \ln(u) + u - \left(\frac{u^2}{2.2;} + \frac{u^3}{3.3;} - \frac{u^4}{4.4;} + \ldots \right)$$

Theis Analytical Solution

As mentioned already, Theis analytical solution was based on the analogy between groundwater flow and heat conduction. In case of heat conduction, the change in temperature (v) at a point $p(x,y)$ at any time t due to an instantaneous line source (x) coinciding with the Z axis can be obtained using the following equation given by Carslaw (1921).

$$v(x,y,t) = \frac{x}{4\pi kt} e^{-\left(x^2+y^2\right)/4kt}$$

Here, k is the Kelvin›s coefficient of diffusivity

For continuous source or sink $x(\tau)$

$$v(x,y,t) = \int_0^t \frac{x(\tau)}{4\pi k(t-t')} e^{-\left(x^2+y^2\right)/4k(t-\tau)} dr$$

For constant source $x(\tau)=x$

$$v(x,y,t) = \frac{x}{4\pi k} \int_0^t \left[\frac{e^{-\left(x^2+y^2\right)/4k(t-\tau)}}{(t-\tau)} \right] d\tau$$

Considering

$$u = \frac{x^2 + y^2}{4K(t-\tau)}$$

When,

$$\tau = 0, \qquad\qquad u = \frac{x^2 + y^2}{4Kt}$$

$$\tau = t, \qquad\qquad u = \infty$$

and $$\qquad\qquad d\tau = \frac{x^2 + y^2}{4K} \frac{1}{u^2} du$$

Then,

$$v(x,y,t) = \frac{x}{4\pi k} \int_{\frac{x^2+y^2}{4Kt}}^{\infty} \left[\frac{e^{-u}}{(t-\tau)} \right] \frac{x^2 + y^2}{4K} \frac{1}{u^2} du$$

$$v(x,y,t) = \frac{x}{4\pi k} \int_{\frac{x^2+y^2}{4Kt}}^{\infty} \left[\frac{e^{-u}}{u^2} \right] u \, du$$

$$v(x,y,t) = \frac{x}{4\pi k} \int_{\frac{x^2+y^2}{4Kt}}^{\infty} \frac{e^{-u}}{u} du$$

Equation above derived for calculation of change in temperature can also be applied for calculation of drawdown at any point (x,y) at any time t. The coefficient of diffusivity is analogous to the coefficient of transmissivity of the aquifer divided by the specific storage (S_s) of the aquifer. The continuous strength of the source and sink is analogous to the discharge rate divided by the specific storage. The equation above in case of drawdown in confined aquifer can be written as

$$s(x,y,t) = \frac{Q/S_S}{4\pi T / S_S} \int_{\frac{x^2+y^2}{4(T/S)t}}^{\infty} \frac{e^{-u}}{u} du$$

$$s(x,y,t) = \frac{Q}{4\pi T} \int_{\frac{S_s(x^2+y^2)}{4Tt}}^{\infty} \frac{e^{-u}}{u} du$$

Putting $x^2 + y^2 = r^2$

$$s(r,t) = \frac{Q}{4\pi T} \int_{\frac{S_s r^2}{4Tt}}^{\infty} \frac{e^{-u}}{u} du$$

Equation above can be used to calculate the drawdown at a distance of r at any time t when water is pumped at a constant rate of Q from the well. This solution is valid homogeneous isotropic aquifer having infinite areal extent and uniform thickness.

Alternate Analytical Solution of radial Flow Equation

The flow equation we have derived early

$$\frac{\partial^2 h}{\partial r^2} + \frac{1}{r}\frac{\partial h}{\partial r} = \frac{S_s}{T}\frac{\partial h}{\partial t}$$

Let us consider

$$u = \frac{r^2 S_s}{4Tt}$$

Thus,

$$r = \sqrt{\frac{4Ttu}{S}}$$

And

$$t = \frac{r^2 S_s}{4Tu}$$

Now we can write that

$$\frac{\partial h}{\partial r} = \frac{\partial h}{\partial u}\frac{\partial u}{\partial r} = \frac{\partial h}{\partial u}\left(\frac{2rS_s}{4Tt}\right) = \frac{\partial h}{\partial u}\left(\frac{2}{r}\frac{r^2 S_s}{4Tt}\right) = 2\left(\frac{u}{r}\right)\frac{\partial h}{\partial u}$$

We can also write

$$\frac{\partial^2 h}{\partial r^2} = \frac{\partial}{\partial r}\left(\frac{\partial h}{\partial r}\right) = \frac{\partial}{\partial r}\left(2\frac{u}{r}\frac{\partial h}{\partial u}\right)$$

$$\Rightarrow 2\frac{\partial}{\partial r}\left(\frac{u}{r}\right)\frac{\partial h}{\partial u} + 2\frac{u}{r}\frac{\partial}{\partial r}\left(\frac{\partial h}{\partial u}\right)$$

Now as defined earlier in the equation above, we have

$$\frac{u}{r} = \frac{rS_s}{4Tt}$$

$$\frac{\partial}{\partial r}\left(\frac{u}{r}\right) = \frac{\partial}{\partial r}\left(\frac{rS_s}{4Tt}\right) = \frac{S_s}{4Tt} = \frac{u}{r^2}$$

We can also write

$$\frac{\partial}{\partial r}\left(\frac{\partial h}{\partial u}\right) = \frac{\partial}{\partial u}\left(\frac{\partial h}{\partial u}\right)\frac{\partial u}{\partial r}$$

From equation above

$$\frac{\partial u}{\partial r} = \frac{2rS_s}{4Tt} = \frac{2u}{r}$$

Putting the equation

$$\frac{\partial}{\partial r}\left(\frac{\partial h}{\partial u}\right) = \left(\frac{\partial^2 h}{\partial u^2}\right)\frac{2u}{r}$$

Now putting the above equation

$$\frac{\partial^2 h}{\partial r^2} = 2\frac{u}{r^2}\frac{\partial h}{\partial u} + 2\frac{u}{r}\frac{2u}{r}\frac{\partial^2 h}{\partial u^2}$$

$$\Rightarrow \frac{\partial^2 h}{\partial r^2} = 2\frac{u}{r^2}\frac{\partial h}{\partial u} + 4\frac{u}{r}\frac{\partial^2 h}{\partial u^2}$$

Now,

$$\frac{\partial h}{\partial t} = \frac{\partial h}{\partial u}\frac{\partial u}{\partial t}$$

From the equation above,

$$\frac{\partial u}{\partial t} = \frac{\partial}{\partial t}\left(\frac{r^2 S_s}{4Tt}\right) = -\frac{r^2 S_s}{4Tt^2} = -\frac{r^2 S_s}{4T}\frac{4^2 T^2 u^2}{r^4 S_s^2} = -4\frac{T}{S_s}\frac{u^2}{r^2}$$

Putting in the equations above,

$$\frac{\partial h}{\partial t} = -4\frac{T}{S_s}\frac{u^2}{r^2}\frac{\partial h}{\partial u}$$

Putting the equations together,

$$2\frac{u}{r^2}\frac{\partial h}{\partial u} + 4\frac{u^2}{r^2}\frac{\partial^2 h}{\partial u^2} + \frac{1}{r}2\left(\frac{u}{r}\right)\frac{\partial h}{\partial u} = -\frac{S_s}{T}4\frac{T}{S_s}\frac{u^2}{r^2}\frac{\partial h}{\partial u}$$

$$4\frac{u}{r^2}\frac{\partial h}{\partial u} + 4\frac{u^2}{r^2}\frac{\partial^2 h}{\partial u^2} = -4\frac{u^2}{r^2}\frac{\partial h}{\partial u}$$

$$\Rightarrow \frac{1}{u}\frac{\partial h}{\partial u} + \frac{\partial^2 h}{\partial u^2} = -\frac{\partial h}{\partial u}$$

$$\Rightarrow \frac{\partial^2 h}{\partial u^2} + \left(\frac{1}{u}+1\right)\frac{\partial h}{\partial u} = 0$$

Taking $f(u) = \frac{\partial h}{\partial u}$

$$\Rightarrow \frac{\partial f}{\partial u} + \left(\frac{1}{u}+1\right)f = 0$$

The Solution of the Differential Equation is

$$f(u) = \frac{\partial h}{\partial u} = Ce^{(-\ln u - u)}$$

$$= \frac{c}{u}e^{-u}$$

Where C is a constant

$$\int_{h(u)}^{h(\infty)} \partial h = C\int_{u}^{\infty} \frac{e^{-u}}{u} \partial u$$

$$= CW(u)$$

$$h(\infty) - h(u) = cw(u)$$

Now next step is to find the value of the constant C.

As per Darcy's law at the well face, the discharge from the well is

$$Q = KA\frac{\partial h}{\partial r}$$

$$\Rightarrow Q = K2\pi r_w b\frac{\partial h}{\partial r}$$

Where r_w is the radius of the well and b is the thickness of the confined aquifer.

From equation above, we have

$$\frac{\partial h}{\partial r} = 2\left(\frac{u}{r_w}\right)\frac{\partial h}{\partial u}$$

Putting in equation above, we have

$$Q = K2\pi r_w b2\left(\frac{u}{r_w}\right)\frac{\partial h}{\partial u}$$

$$Q = 4\pi Tu\frac{\partial h}{\partial u}$$

Putting equation above, we have

$$Q = 4\pi Tu\frac{c}{u}e^{-u}$$

$$C = \frac{Q}{4\pi T \, e^{-u}} = \frac{Q}{4\pi T} e^{u}$$

For most of the well the radius of the well is very small and the $e^{u} = 1$.

As such the constant C can be written as

$$C = \frac{Q}{4\pi T}$$

Putting the value of C in the equations we have

$$h(\infty) - h(u) \frac{Q}{4\pi T} W(u)$$

$$Or, \, s(r,t) = \frac{Q}{4\pi T} W(u)$$

The equation above can be used to calculate the drawdown at a distance of r at a time t when water is pumped at a constant rate of Q from the well. This solution is valid for homogeneous isotropic aquifer having infinite areal extent and uniform thickness.

Well Test

A well test is conducted to evaluate the amount of water that can be pumped from a particular water well. More specifically, a well test will allow prediction of the maximum rate at which water can be pumped from a well, and the distance that the water level in the well will fall for a given pumping rate and duration of pumping.

Well testing differs from aquifer testing in that the behaviour of the well is primarily of concern in the former, while the characteristics of the aquifer (the geological formation or unit that supplies water to the well) are quantified in the latter.

When water is pumped from a well the water level in the well falls. This fall is called drawdown. The amount of water that can be pumped is limited by the drawdown produced. Typically, drawdown also increases with the length of time that the pumping continues.

Well Losses vs. Aquifer Losses

The components of observed drawdown in a pumping well were first described by Jacob (1947), and the test was refined independently by Hantush (1964) and Bierschenk (1963) as consisting of two related components,

$$s = BQ + CQ^{2},$$

where s is drawdown (units of length e.g., m), Q is the pumping rate (units of volume flowrate e.g., m³/day), B is the aquifer loss coefficient (which increases with time — as predicted by the Theis solution) and C is the well loss coefficient (which is constant for a given flow rate).

The first term of the equation (BQ) describes the linear component of the drawdown; i.e., the part in which doubling the pumping rate doubles the drawdown.

The second term (CQ^2) describes what is often called the 'well losses'; the non-linear component of the drawdown. To quantify this it is necessary to pump the well at several different flow rates (commonly called *steps*). Rorabaugh (1953) added to this analysis by making the exponent an arbitrary power (usually between 1.5 and 3.5).

To analyze this equation, both sides are divided by the discharge rate (Q), leaving s/Q on the left side, which is commonly referred to as *specific drawdown*. The right hand side of the equation becomes that of a straight line. Plotting the specific drawdown after a set amount of time ($Ät$) since the beginning of each step of the test (since drawdown will continue to increase with time) versus pumping rate should produce a straight line.

$$\frac{s}{Q} = B + CQ$$

Fitting a straight line through the observed data, the slope of the best fit line will be C (well losses) and the intercept of this line with $Q = 0$ will be B (aquifer losses). This process is fitting an idealized model to real world data, and seeing what parameters in the model make it fit reality best. The assumption is then made that these fitted parameters best represent reality (given the assumptions that went into the model are true).

The relationship above is for fully penetrating wells in confined aquifers (the same assumptions used in the Theis solution for determining aquifer characteristics in an aquifer test).

Well Efficiency

Often the *well efficiency* is determined from this sort of test, this is a percentage indicating the fraction of total observed drawdown in a pumping well which is due to aquifer losses (as opposed to being due to flow through the well screen and inside the borehole). A perfectly efficient well, with perfect well screen and where the water flows inside the well in a frictionless manner would have 100% efficiency. Unfortunately well efficiency is hard to compare between wells because it depends on the characteristics of the aquifer too (the same amount of well losses compared to a more transmissive aquifer would give a lower efficiency).

Specific Capacity

Specific capacity is a quantity that which a water well can produce per unit of drawdown. It is normally obtained from a step drawdown test. Specific capacity is expressed as:

$$S_c = \frac{Q}{h_0 - h}$$

where

S_c is the specific capacity ($[L^2T^{-1}]$; m²/day or USgal/day/ft)

Q is the pumping rate ($[L^3T^{-1}]$; m³/day or USgal/day), and

$h_0 - h$ is the drawdown ($[L]$; m or ft)

The specific capacity of a well is also a function of the pumping rate it is determined at. Due to non-linear well losses the specific capacity will decrease with higher pumping rates. This complication makes the absolute value of specific capacity of little use; though it is useful for comparing the efficiency of the same well through time (e.g., to see if the well requires rehabilitation).

This section will deal with the solution of unsteady flow problem in case of unconfined aquifer and also in case of leaky confined aquifer.

Unconfined Aquifer

The time drawdown relationship is complex in case of unconfined aquifer. When drawdown is small compare to the saturated thickness of the aquifer, the solution method applied for the case of confined aquifer can also be applied to the case of unconfined aquifer. However, when drawdown is significant, the solution method applied to the case of confined aquifer will not be applicable as it will violate the assumptions that has been made in case of confined aquifer. In this case, the water released from the storage will not be discharged instantaneously with the declination of the hydraulic head. In case of unconfined aquifer, in the initial period after the start of the pumping, water is released instantaneously from the storage. This situation is similar to the time drawdown relationship for the case of confined aquifer and can be approximated using the Theis type curve. After some period of time from the start of the pumping, the rate of drawdown will be slow due to the gravity drainage replenishment from the pores of the unsaturated zone. The gravity drainage of water from the unsaturated zone proceeds in a variable rate. Finally, an equilibrium condition is achieved between gravity drainage and rate of decline of water table at later stage. This portion of the time drawdown relationship can also be approximate with the Theis type curve. Neuman (1975) gave the following solution for unconfined aquifer with fully penetrated well and constant discharge considering delayed yield.

$$S = \frac{q}{4\pi t} w\left(u, u_y, \eta\right)$$

Where

$$u = \frac{S_s r^2}{4Tt}$$

$$u_y = \frac{S_y r^2}{4Tt}$$

$$\eta = \frac{r^2 K_z}{b^2 K_h}$$

It may be noted that u is applicable for the early drawdown data whereas u_y is applicable for later drawdown data. $W(u, u_y, \eta)$ can be calculated using the curve generated by Neuman (1975).

Wells in a Leaky Confined Aquifer

A confined aquifer will be called a leaky aquifer when water is withdrawn from the confined aquifer, there is a vertical flow from the overlaying aquitard as shown in figure below. After the starts of the pumping, the lowering of piezometric head in the aquifer builds hydraulic gradient within the aquitard. As a result of the hydraulic gradient, downward vertical groundwater flow takes place through the aquitard.

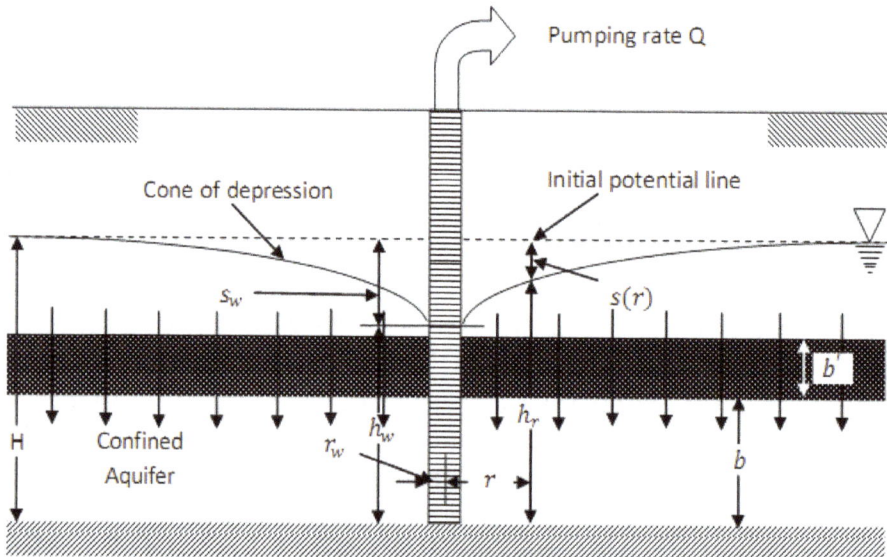

A leaky confined aquifer

The drawdown of the piezometric surface can be obtained by (Hantush 1956, Cobb *et al.* 1982)

$$s = \frac{Q}{4\pi T} W\left(u, \frac{r}{B}\right)$$

Where

$$W\left(u, \frac{r}{B}\right) = \int_0^\infty \exp\{-u - r^3 / [4B^2(x+u)]\} / (x+u)\exp(-x)dx$$

$$u = \frac{S_s r^2}{4Tt}$$

$$\frac{r}{B} = \frac{r}{\sqrt{T/(K'/b')}}$$

Where T is the transmissivity of the leaky confined aquifer, K' is the vertical hydraulic conductivity of the aquitard, and b' is the thickness of the aquitard.

Partially Penetrating Well

In a well when the intake of the well is less than the thickness of the well, then the well is called partially penetrated well. In case of partially penetrated well, the flow lines are not truly horizontal near the well. The flow lines are curved upward or downward near the well. However, at a distance far away from the well, the flow lines are horizontal. As a result of non-horizontal nature of the flow lines near the well, the length of the flow lines are more than the case of a fully penetrated well. Thus the drawdown in case of partially penetrating well is more than the fully penetrating well. Figure below shows a partially penetrated well.

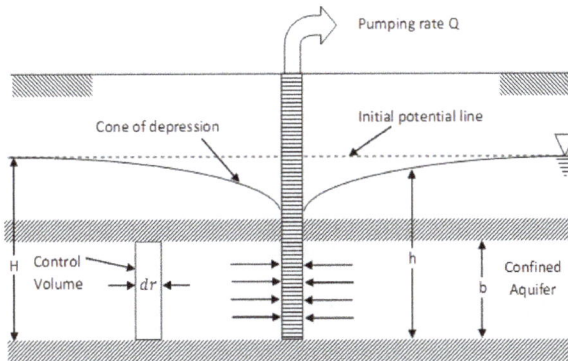

Partially penetrated well

The drawdown of the partially penetrated well can be written as

$$s_p = s + \Delta s$$

Where, S is the drawdown of the fully penetrated well and Δ_s is the additional drawdown due to partial penetration.

For the Figure given below,

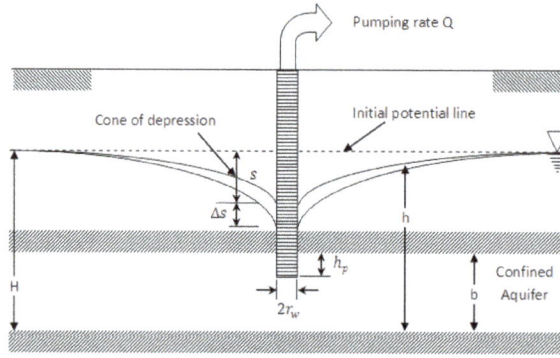

Partially penetrated well

The additional drawdown, Δ_s can be calculated as (Todd and Mays, 2011)

$$\Delta s = \frac{Q}{2\pi T}\frac{1-p}{p}\ln\left(\frac{(1-p)h_p}{r_w}\right)$$

Change in Hydraulic Properties Near a Well

Consider a case of a pumping well as shown in Fig. below.

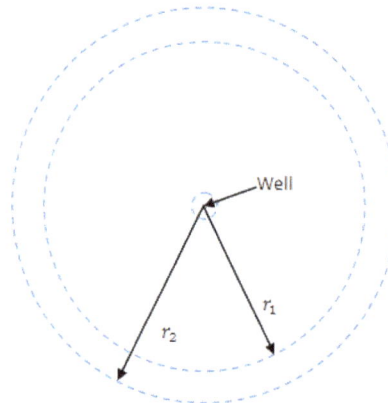

Pumping well

The discharge of the well can be expressed as

$$Q = A_1V_1 = A_2V_2$$

$$\Rightarrow 2\pi r_1 h V_1 = 2\pi r_2 h V_2$$

$$\Rightarrow r_1V_1 = r_2V_2$$

Here $r_2 > r_1$

So $V_1 > V_2$

Therefore, velocity near the well is more than the velocity away from the well. Due to the high velocity in the vicinity of the well , the fine particles that are present in the aquifer formation are moved with the flow of water. As a result of this phenomenon, the permeability of the aquifer medium will be more in the vicinity of the well.

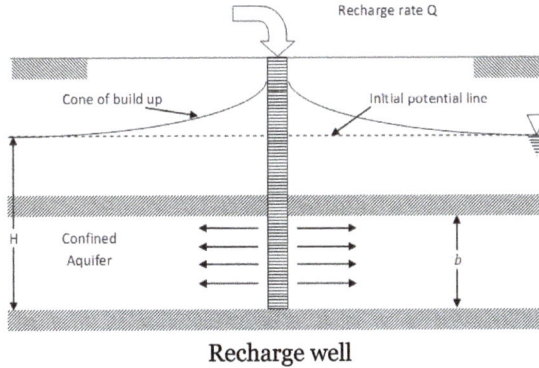

Recharge well

Now, in case of recharge well figure above, the impurities that are present in water are also move along with the injected water to the aquifer medium. As the velocity of flow in the vicinity of the well is higher, the impurities present in the water will move along with the water and will settle down at some distance from the well. As a result of the settlement of impurities, the permeability of the medium will reduce. As such the reduction on permeability should be considered in modeling the flow in an aquifer due to artificial recharge.

Multiple Well Systems

In a well field, when cone of depression of one well overlaps with the cone of depression of other wells, then the actual drawdown will be more than the drawdown calculated for the individual well figure (Partially penetrated well). In this case, the actual drawdown can be calculated using the principle of superposition of linear system.

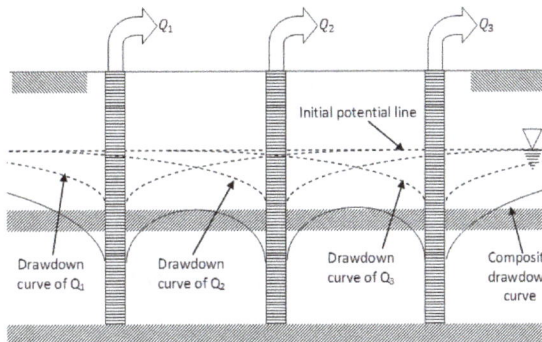

Multiple well system

For a well field of n wells, the actual drawdown can be calculated as

$$s_a(r,t) = s_1(r,t) + s_2(r,t) + s_3(r,t) + s_4(r,t) + s_5(r,t) + \ldots + s_n(r,t)$$

or, $s_a(r,t) = \sum_{i=1}^{n} s_i(r,t)$

Where S_a is the actual drawdown at a distance r at time t, S_i is the drawdown at that point caused by the discharge of the well i at time t, n is the number of wells in the well fields.

Figure below explains the interference of cone of depression of two pumping wells. The coordinates of the two wells. The individual cone of depression of the two wells are shown on the Fig. below. The combine effect of the two wells can be obtained by adding the individual drawdown of the two wells, *i.e.* if drawdown of the first well is S_1 and the second well is S_2, the combine drawdown will be $S = S_1 + S_2$. The combine effect is shown in Fig.

Drawdown of first well

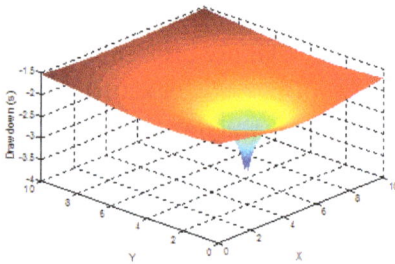

Drawdown of second well

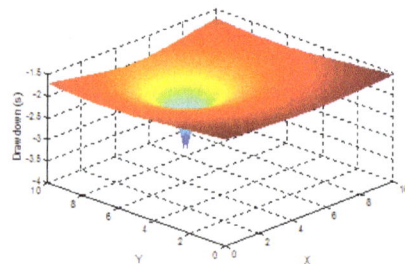

Combine drawdown

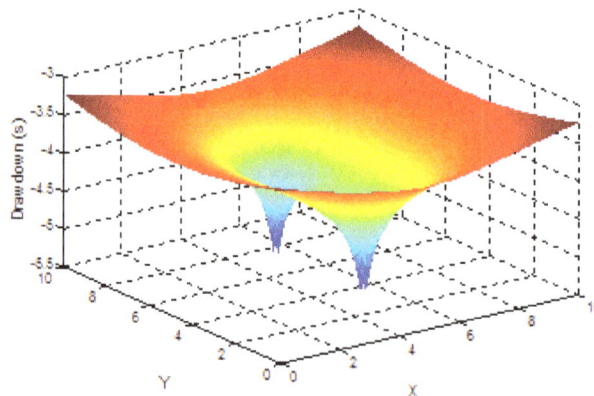

Cone of depression of multiple wells system

Wells Near Aquifer Boundaries

The assumption of infinite horizontal extend is no longer valid when water is pumped from a well near the aquifer boundary. Method of superposition can be used to implement the effect of aquifer boundary by adding a well at different location. The well that creates the same effect as boundary is called image well.

Well Near a Stream

Figure below shows a well near a stream. In this case, the actual drawdown at the stream boundary will be zero as stream is considered as an infinite source. In order to maintain zero drawdown, an imaginary recharge well is considered at a distance equal to the distance between the pumping well and the stream boundary.

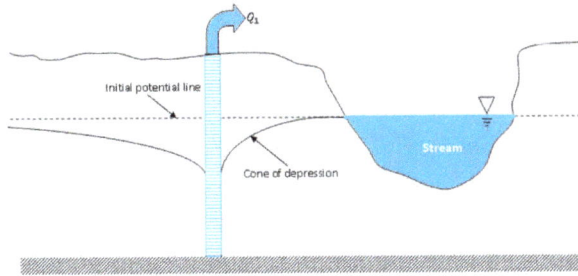

Well near a stream

Fig. (Recharge well) shows an equivalent hydraulic system in an aquifer of infinite areal extend. For the equivalent hydraulic system, the time drawdown relationship for the pumping well and also for the imagery recharge well can be obtained separately. The actual drawdown can be obtained using the principle of superposition.

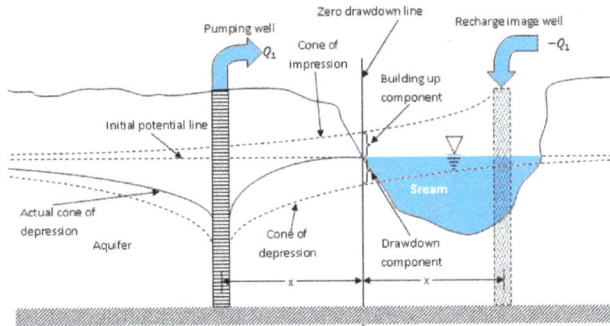

Equivalent hydraulic system in a aquifer of infinite areal extend

Consider the figure below. The pumping well is at a distance of x from the stream boundary. In order to calculate the actual drawdown at the observation location, an image well is

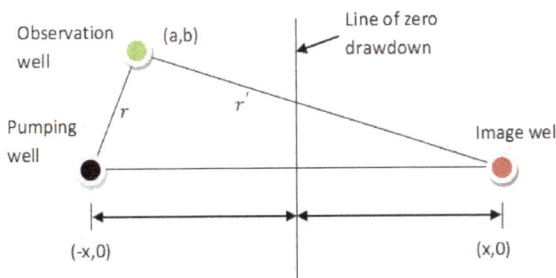

Pumping well, Observation well and Image well

considered at a distance of x on the other side of the line of zero drawdown. The distance of the observation well from the pumping well is r and from the image well is r'.

For the steady state condition of a confined aquifer, the drawdown at the observation well can be obtained as

$$s(a,b) = \frac{Q}{2\pi T} \ln\left(\frac{R}{r}\right) + \frac{-Q}{2\pi T} \ln\left(\frac{R}{r'}\right)$$

$$\Rightarrow s(a,b) = \frac{Q}{2\pi T} \ln\left(\frac{r'}{r}\right)$$

$$\Rightarrow s(a,b) = \frac{Q}{4\pi T} \ln\left(\frac{(a+x)^2 + b^2}{(a-x)^2 + b^2}\right)$$

For the unsteady condition, the drawdown at r at any time t can be obtained as

$$s(r,t) = \frac{Q}{4\pi T} W\left(\frac{r^2 S_s}{4Tt}\right) + \frac{-Q}{4\pi T} W\left(\frac{r'^2 S_s}{4Tt}\right)$$

$$s(r,t) = \frac{Q}{4\pi T} \left[W\left(\frac{r^2 S_s}{4Tt}\right) - W\left(\frac{r'^2 S_s}{4Tt}\right) \right]$$

Well Near an Impermeable Boundary

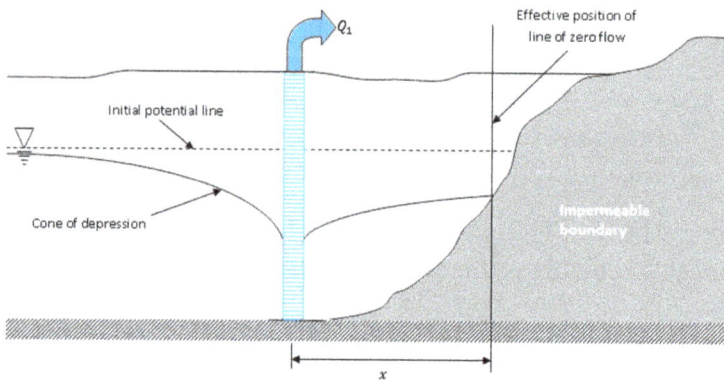

Well near an impermeable boundary

Figure above shows a well near an impermeable boundary. In this case, the actual drawdown at the impermeable boundary will be more than the drawdown calculated considering infinite areal extend of the aquifer medium. This problem can be solved by considering an imaginary pumping well at a distance equal to the distance between the pumping well and the image pumping well. Figure below has shown the equivalent hydraulic system in an aquifer with infinite areal extent. For the equivalent hydraulic system, the time drawdown relationship for the pumping well and also for the imagery recharge well can be obtained separately. The actual drawdown can be obtained using the principle of superposition.

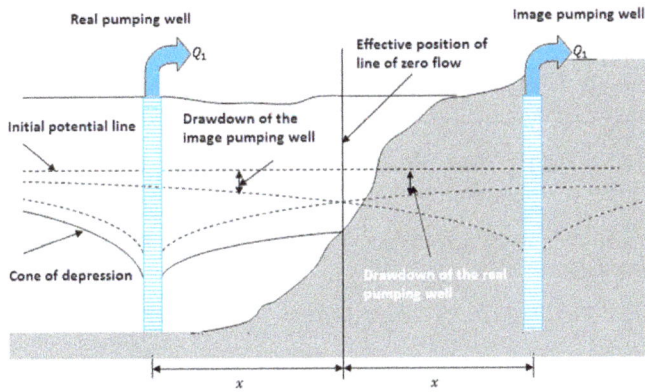

Equivalent hydraulic system in a aquifer of infinite areal extend

Consider the figure below. The pumping well is at a distance of x from the impermeable boundary. In order to calculate the actual drawdown at the observation location, an image well is considered at a distance of x on the other side of the line of zero flow. The distance of the observation well from the pumping well is r and from the image well is r'.

For the unsteady condition, the drawdown at a distance r at any time t can be obtained as,

$$s(r,t) = \frac{Q}{4\pi T} W\left(\frac{r^2 S_s}{4Tt}\right) + \frac{Q}{4\pi T} W\left(\frac{r'^2 S_s}{4Tt}\right)$$

$$s(r,t) = \frac{Q}{4\pi T}\left[W\left(\frac{r^2 S_s}{4Tt}\right) + W\left(\frac{r'^2 S_s}{4Tt}\right)\right]$$

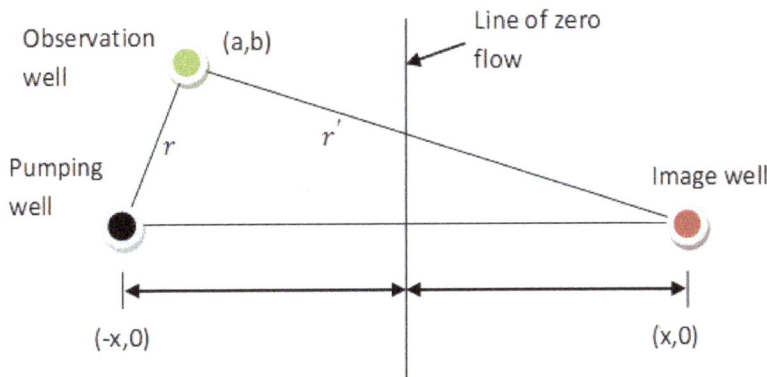

Pumping well, observation well and image well

References

- Eckert, Michael (2006). The Dawn of Fluid Dynamics: A Discipline Between Science and Technology. Wiley. p. ix. ISBN 3-527-40513-5

- Severn, Richard. "Environment Agency Field Test Report – TIENet 360 LaserFlow" (PDF). RS Hydro. RS Hydro-Environment Agency. Retrieved 3 August 2015

- Furness, Richard A. (1989). Fluid flow measurement. Harlow: Longman in association with the Institute of Measurement and Control. p. 21. ISBN 0582031656

- Miller, Richard W. (1996). Flow Measurement Engineering Handbook (3rd ed.). Mcgraw Hill. p. 6.16–6.18. ISBN 0070423660

- Adrian, R. J., editor (1993); Selected on Laser Doppler Velocimetry, S.P.I.E. Milestone Series, ISBN 978-0-8194-1297-3

4

Preservation of Groundwater

It is necessary to preserve aquifers. Here, groundwater management takes importance. It analyzes and predicts the condition of aquifers and aims to achieve a sustainable approach towards use of water. A few of its objectives are that it maximizes water withdrawal; it minimizes the energy required to extract and distribute water. This chapter has been carefully written to provide an easy understanding of the varied facets of groundwater preservation.

Groundwater Resources Management

Groundwater management aims to achieve certain strategies for sustainable use of the resources. Some of the objectives of the groundwater management models are:

1. Maximization of total withdrawal from an aquifer.

2. Minimization of cost of per unit volume of water supplied to the consumers.

3. Maximization of net benefit of a project related to the supply of water for municipal, industrial, agricultural, *etc.* purposes.

4. Minimization of total consumption of energy of project related to the extraction and distribution of groundwater.

5. Minimization of an error function obtained from the sum of the absolute difference between actual and predicted values of certain water level.

In achieving these objectives, certain restrictions have to be satisfied for obtaining physically meaningful optimal policies. These restrictions are known as constraints. Some of the commonly applicable constraints in groundwater management models are:

1. Water level at any location in an aquifer should not go below certain value. This constraint will put restriction on the depletion of water table due to pumping of water from an aquifer.

2. Water level at any location in an aquifer should not rise above certain level. This is required to avoid water logging of an area and also to dewater an area under construction.

3. Limits on the maximum and minimum withdrawal of water from an aquifer.

4. Spring discharge should not go below certain specified value.

5. Base flow of a stream or a river should not go below certain minimum value.

6. Quality of pumped water should not deteriorate below certain specific value.

7. Limit on land subsidence.

Solution Techniques of Groundwater Management Models

A groundwater management model should include the groundwater simulation model as constraint along with other managerial constraints. Incorporation of groundwater simulation model ensures that the obtained management strategies are physically feasible. The simulation model simulates the physical processes of the aquifers. Generally embedding technique and response matrix approach (Gorelick, 1983) are used for incorporating the governing equations within the management model. The embedded optimization technique incorporates finite difference or finite element approximation of the governing equations as equality constraints within the management model, along with the other physical and managerial constraints. Some of the application of embedding technique for groundwater management problems are seen in Das, 1995; Das and Datta 1999, etc. However, this approach is not suitable for large aquifer systems. The approach may be numerically inefficient especially when applied to large aquifer systems with considerable heterogeneity. The response matrix approach is based on the principle of superposition and linearity. The performance of response matrix approach is not suitable for highly nonlinear systems (Rosenwald and Green, 1974).

As an alternative to the embedding technique and the response matrix approach, the simulation model may be incorporated with the management model as an external module. In this approach, an external simulation model is linked to the optimization model (Finney at el., 1992; Emch and Yeh, 1998; Bhattacharjya and Datta, 2009). The optimization model calls the simulation model as and when it requires any information from the simulation model. The methodology has been applied effectively for large scale groundwater management models. The main disadvantage of this approach is that numerous repetitive iterations between the simulation model and the optimizer are required to arrive at an optimal solution. The computational time can be substantially reduced by utilizing parallel processing capabilities of advanced computers. This would enable use of rigorous numerical models for simulation and its linkage to an optimization model. Also the time requirement for iterative solutions of the optimization model and the simulation model can be drastically reduced. However, this requires appropriate computer hardware and numerical simulation models specially tailored to explicit parallel processing capabilities.

An optimization technique has to be used for solving the management model. Classical optimization techniques have been applied for solving groundwater management problems. Most of the classical optimization methods use gradient search technique for finding the optimal solution. The performance of the gradient based classical optimization methods is not satisfactory when response surface is highly irregular. In such a situation, it is very likely that the solutions obtained would be local optimal solutions. One possible remedy is the use multiple solution points as initial solutions. Some people have also used global search techniques, such as Genetic Algorithm, Simulated Annealing, Differential evolution, *etc.* for solving groundwater management models.

Solution of Simple Groundwater Management Model

(a) Confined Aquifer

Now we will solve a very simple groundwater management problem. Consider a case of one dimensional steady state flow in a confined aquifer. The governing equation in this case can be written as,

$$T_x \frac{\partial^2 h}{\partial x^2} + N(x) = 0$$

The finite difference form of the equation can be written as,

$$\frac{h_{i+1} - 2h_i + h_{i-1}}{(\Delta x)^2} + \frac{N_i}{Tx} = 0$$

$$h_{i-1} - 2h_i + h_{i-1} + \frac{(\Delta x)^2}{Tx} N_i = 0$$

$$h_{i=1} - 2h_i + h_{i+1} + PN_i = 0 \qquad Considering \ P = \frac{(\Delta x)^2}{Tx}$$

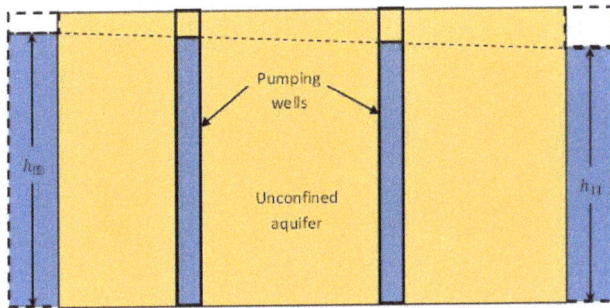

(a)

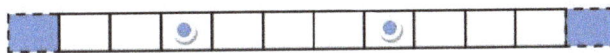

(b)

(a) Confined aquifer (b) Top view of the aquifer

Consider the confined aquifer as shown in above figure. The aquifer has been discretized to 10 blocks. Constant head boundary is considered on both upstream and downstream sides of the aquifer. Let the constant head at the upstream side be h_o and constant head at the downstream side of the aquifer be h_{11}. Let us consider that the heads at the block centers are h_1 to h_{10} as shown in the figure below. There are two pumping wells in the aquifer as shown in the figure. Let us also consider that the main objective of the management model is to withdrawal an amount of water equal to N_{min}. The finite difference form of the governing equation at each block center has to be incorporated as constraint with the optimization problem along with the other constraints.

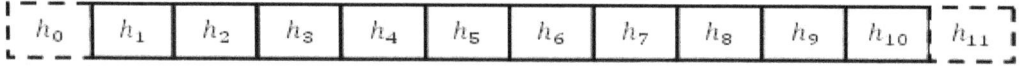

Discretized aquifer

Inclusion of these constraints ensures that the solution obtained would be physically feasible. The optimization problem can be formulated as,

Maximize $\int = \sum_{k=1}^{10} h_k$

Subject to

$$-2h_1 + h_2 + PN_1 = h_0$$

$$h_1 - 2h_2 + h_3 + PN_2 = 0$$

$$h_2 - 2h_3 + h_4 + PN_3 = 0$$

$$h_3 - 2h_4 + h_5 + PN_4 = 0$$

$$h_4 - 2h_5 + h_6 + PN_5 = 0$$

$$h_5 - 2h_6 + h_7 + PN_6 = 0$$

$$h_6 - 2h_7 + h_8 + PN_7 = 0$$

$$h_7 - 2h_8 + h_9 + PN_8 = 0$$

$$h_8 - 2h_9 + h_{10} + PN_9 = 0$$

$$h_9 - 2h_{10} + PN_{10} = h_{11}$$

$$\sum_{k=1}^{10} N_k \geq N_{min}$$

$$h_k \geq 0 \quad for \ k = 1, 2 \ldots 10$$

$N_k \geq 0 \quad for \ k = 1,2\ldots10$

The objective function will try to maximize the head value of the system while satisfying all the constraints.

The constraint ensures that the minimum withdrawn is N_{min} and the constraints are the non negativity constraints. Here the decision variables are N_1 to N_{10} and h_1 to h_{10} are the state variables. If we look at the optimization problem, the objective function and the constraints are linear in nature. As such the problem is a linear problem (LP) and can be solved using any LP solving algorithm like Simplex method.

The solution of the optimization problem will give the spatial distribution of pumping pattern where total pumping from all the pumping wells is equal to N_{min}.

(b) Unconfined Aquifer

Now consider a case of one dimensional steady state flow in an unconfined aquifer. The governing equation in this case can be written as,

$$\frac{\partial}{\partial x}\left(T_x \frac{\partial h}{\partial x}\right) + N(x) = 0$$

Considering $T_x = Kh$

$$\frac{\partial^2 h^2}{\partial x^2} + \frac{2N(x)}{K} = 0$$

The above equation is a nonlinear equation. The equation can be made linear by considering $d = h^2$. Thus the equation can be written as,

$$\frac{\partial^2 d}{\partial x^2} + \frac{2N(x)}{K} = 0$$

The finite difference form of the equation can be written as,

$$\frac{d_{i+1} - 2d_i + d_{i-1}}{(\Delta x)^2} + \frac{2N_i}{K} = 0$$

$$d_{i-1} - 2d_i + d_{i+1} + \frac{2(\Delta x)^2}{K}N_i = 0$$

$$d_{i-1} - 2d_i + d_{i+1} + RN_i = 0 \quad Considering \ R = \frac{2(\Delta x)^2}{K}$$

(a)

(b)

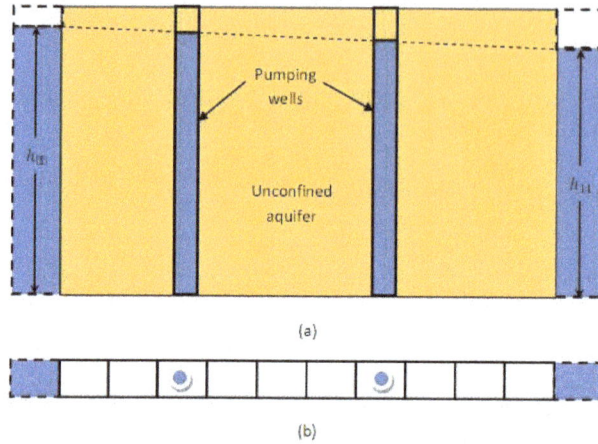

(a) Confined aquifer (b) Top view of the aquifer

Consider the unconfined aquifer as shown in the above figure. The aquifer has been discretized to 10 blocks. Constant head boundary is considered on both upstream and downstream sides of the aquifer. Lets the constant head at the upstream side is h_0 and constant head at the downstream side of the aquifer is h_{11}. Let us consider that the heads at the block centers are h_1 to h_{10} as shown in the figure below.

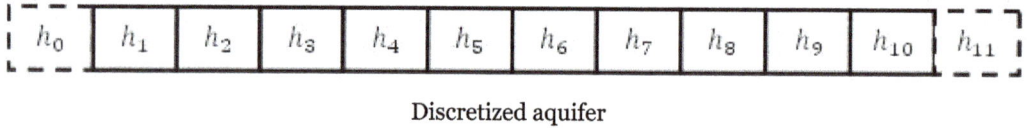

Discretized aquifer

There are two pumping wells in the aquifer as shown in the figure ((a) Confined aquifer (b) Top view of the aquifer). Let us also consider that the main objective of the management model is to withdraw an amount of water equal to N_{min}. The finite difference form of the governing equation at each block centers has to be incorporated as constraints with the optimization problem along with the other constraints. Inclusion of these constraints ensures that the solution obtained would be physically feasible. The optimization problem can be formulated as,

Maximize $f = \sum_{k=1}^{10} d_k$

Subject to

$$-2d_1 + d_2 + RN_1 = d_0$$

$$d_1 - 2d_2 + d_3 + RN_2 = 0$$

$$d_2 - 2d_3 + d_4 + RN_3 = 0$$

$$d_3 - 2d_4 + d_5 + RN_4 = 0$$

$$d_4 - 2d_5 + d_6 + RN_5 = 0$$

$$d_5 - 2d_6 + d_7 + RN_6 = 0$$

$$d_6 - 2d_7 + d_8 + RN_7 = 0$$

$$d_7 - 2d_8 + d_9 + RN_8 = 0$$

$$d_8 - 2d_9 + d_{10} + RN_9 = 0$$

$$d_9 - 2d_{10} + RN_{10} = d_{11}$$

$$\sum_{k=1}^{10} N_k \geq N_{min}$$

$$d_k \geq 0 \quad for \ k = 1, 2 \ldots 10$$

$$N_k \geq 0 \quad for \ k = 1, 2 \ldots 10$$

The constraint ensures that the minimum withdrawn is N_{min} and the constraints and are the non negativity constraints. Here the decision variables are N_1 to N_{10} and h_1 to h_{10} are the state variables. The solution of the optimization problem will give spatial distribution of pumping pattern where total pumping from all the wells will be equal to N_{min}.

Simulation of Flow Process using Finite Difference Method

This lesson will be dealt with the solution of more complicated groundwater management models. Consider case of transient flow in a confined aquifer. The primary design objective is to maximize the amount of pumping from the well field. At the same time, it is also necessary to see that the groundwater table does not deplete beyond certain depth. This problem can be solved using an optimization technique. In order to obtain meaningful groundwater management strategies, the aquifer simulation model is required to incorporate with the optimization model. In this lesson, we will simulate the aquifer flow process using finite difference method.

The transient simulation model of an aquifer gives the head values at each aquifer location after every time interval. The number of such intervals is user defined. For a particular time interval, the pumping rate is constant. The 2D governing equation for this case is,

$$\frac{\partial}{\partial x}\left[T_x \frac{\partial h}{\partial x}\right] + \frac{\partial}{\partial y}\left[T_y \frac{\partial h}{\partial y}\right] + N(x, y) = S \frac{\partial h}{\partial t}$$

Finite difference form of the equation is given as :

$$A_{ij}\left(h_{i+1,j}^{n-1} + h_{i+1,j}^{n}\right) + B_{ij}\left(h_{i-1,j}^{n-1} + h_{i-1,j}^{n}\right) + C_{ij}\left(h_{i,j+1}^{n-1} + h_{i,j+1}^{n}\right) + D_{ij}\left(h_{i,j-1}^{n-1} + h_{i,j-1}^{n}\right) -$$

$$\left(A_{ij} + B_{ij} + C_{ij} + D_{ij} + \frac{2S}{\Delta t}\right)h_{i,j}^{n} + 2N_{i,j}^{n} - \left(A_{ij} + B_{ij} + C_{ij} + D_{ij} + \frac{2S}{\Delta t}\right)h_{i,j}^{n-1} = 0$$

Where,

$$A_{ij} = \frac{1}{(\Delta x)^{2}}\left(\frac{2T_{x}(i+1,j)*2T_{x}(i,j)}{T_{x}(i+1,j) + T_{x}(i,j)}\right)$$

$$B_{ij} = \frac{1}{(\Delta x)^{2}}\left(\frac{2T_{x}(i-1,j)*2T_{x}(i,j)}{T_{x}(i-1,j) + T_{x}(i,j)}\right)$$

$$C_{ij} = \frac{1}{(\Delta y)^{2}}\left(\frac{2T_{y}(i+1,j)*2T_{y}(i,j)}{T_{x}(i+1,j) + T_{y}(i,j)}\right)$$

$$D_{ij} = \frac{1}{(\Delta y)^{2}}\left(\frac{2T_{y}(i-1,j)*2T_{y}(i,j)}{T_{x}(i-1,j) + T_{y}(i,j)}\right)$$

The above equation is rearranged as

$$-\left(A_{ij} + B_{ij} + C_{ij} + D_{ij} + \frac{2S}{\Delta t}\right)h_{i,j}^{n} + A_{ij}h_{i+1,j}^{n} + B_{ij}h_{i-1,j}^{n} + C_{ij}h_{i,j+1}^{n} + D_{ij}h_{i,j-1}^{n} + 2N_{i,j}^{n} =$$

$$\left(A_{ij} + B_{ij} + C_{ij} + D_{ij} + \frac{2S}{\Delta t}\right)h_{i,j}^{n-1} + A_{ij}h_{i+1,j}^{n-1} - B_{ij}h_{i-1,j}^{n-1} + C_{ij}h_{i,j+1}^{n-1} + D_{ij}h_{i,j-1}^{n-1}$$

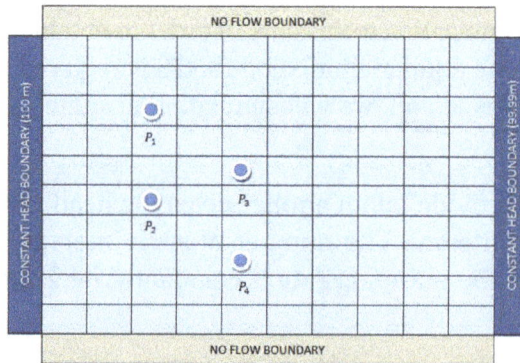

Discretized aquifer

Analyzing the above equation, the left hand side contains the variables involving head values at n^{th} time interval while the right hand side of the equation contains head values at $(n-1)^{th}$ time interval. Hence, knowing the head value at a given time interval, the head values at the next time interval can be calculated, $i.e.$ h^{n+1} can be obtain using h^n, h^{n+2} can be obtain using h^{n+1}, and so on. Therefore, the head distribution at h^o must be known for finding the solution. This is known as initial condition of the aquifer. The h^o can be obtained from field observation. However, for the example problem considered here, the steady state head value is considered as h^o.

Consider the discretized aquifer shown in figure above. The aquifer domain has been divided in nxngrids. There will be total nxn number of equations which is equal to the number of unknowns, $i.e.$ the head value at each node center. The set of nxn equation can be written as,

$$\begin{bmatrix} a_{11} & \cdots & a_{1(nxn)} \\ \vdots & \ddots & \vdots \\ a_{(nxn)1} & \cdots & a_{(nxn)(nxn)} \end{bmatrix} \begin{bmatrix} h_1 \\ \vdots \\ h_{nxn} \end{bmatrix} = \begin{bmatrix} r_1 \\ \vdots \\ r_{nxn} \end{bmatrix}$$

Where a_{11} is the coefficient of h_1 in equation 1. Similarly,$a_{1(nxn)}$ is the coefficient of h_{nxn} in equation, $a_{(nxn)(nxn)}$ is the coefficient of $h_{(nxn)}$ in equation (nxn). r_1 is the right hand side of equation. Similarly, r_{nxn} is the right hand side of equation nxn. The equation above can also be written as,

$AH = R$

where,

$$A = \begin{bmatrix} a_{11} & \cdots & a_{1(nxn)} \\ \vdots & \ddots & \vdots \\ a_{(nxn)1} & \cdots & a_{(nxn)(nxn)} \end{bmatrix}$$

$$H = \begin{bmatrix} h_1 \\ \vdots \\ h_{(nxn)} \end{bmatrix}$$

$$R = \begin{bmatrix} r_1 \\ \vdots \\ r_{nxn} \end{bmatrix}$$

The equation above can be solved as,

$H = A^{-1}R$

Consider the aquifer shown in figure below

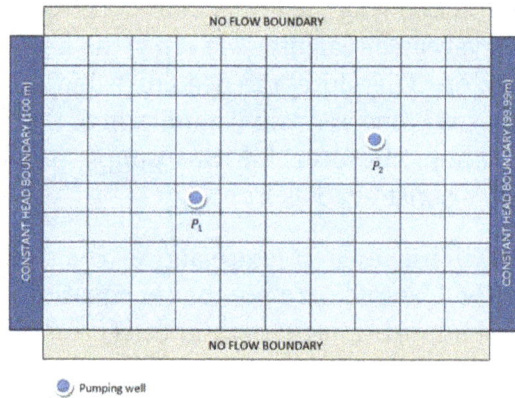

Discretized aquifer with pumping wells

The confined aquifer has constant head of 100.00 m at the left side and constant head of 99.9 m at the right hand. The other two sides are bounded by impermeable layer. Thus no flow boundary

Table: a. Pumping patterns

Time (hr)	Discharge (m³/hr)	
	P_1	P_2
0-4	200	300
5-8	0	200
9-12	300	0
13-16	0	0
17-20	0	0
21-24	300	200

Table: b. Aquifer and simulation parameters

Parameters	Value
Transmissivity (T_x)	300 m²/day
Transmissivity (T_y)	300 m²/day
Storativity	0.001
Δx	100 m
Δy	100 m
Δt	4 h

condition is assumed on these two sides. There are two pumping wells. Pumping patterns of these two pumping wells are shown Table a.

For the aquifer, simulation parameters shown in Table b and the solution obtained using the finite difference method is shown in figure below.

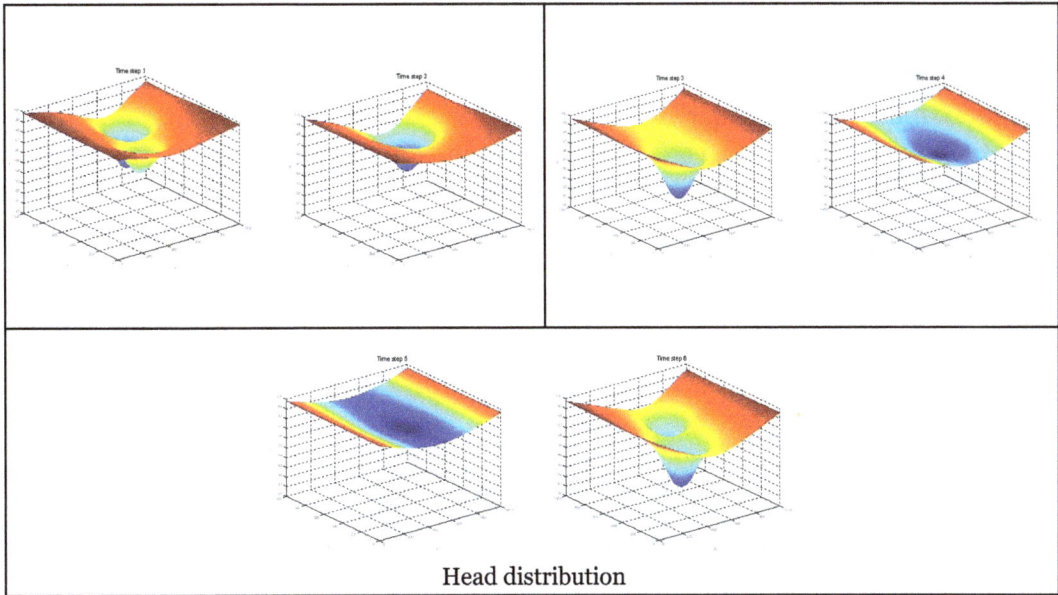

Head distribution

One of the primary objectives of a groundwater management model is to know the maximum possible withdrawn that can be obtained from an aquifer. In this lesson, we will find out the maximum possible withdrawal from a confined aquifer while maintaining drawdown at certain specified level. Consider the aquifer shown in the figure below.

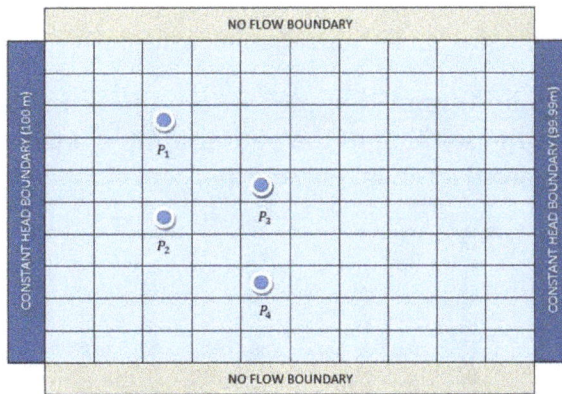

Confined aquifer with pumping well locations

The main objective of the problem is to obtain maximum possible discharge from the pumping wells while maintaining drawdown at certain level. The optimization model can be formulated as,

Minimize $f = \sum_{t=1}^{T} \sum_{n=1}^{N} Q_n^t$

Subject to

$h_{i,j}^t \geq h_{\min}$ $i = 1, 2, \cdots, I; j = 1, 2, \cdots, J, t = 1, 2, \cdots, T$

Where T is the total number of time steps, N is the total number of pumping and Q_i^n is the pumping rate at time t for the well n, l is the total number of grids in x direction, j is the total number of grids in y direction, $h_{i,j}^t$ is the head value at location (i, j) at time t. The management problem can be solved using simulation-optimization methodology. In this methodology, the simulation model is linked externally with the optimization model. In every iteration of the optimization model, the simulation model is called to get the information about spatial and temporal distribution of hydraulic heads in the aquifer. The spatial and temporal distribution of head values is required to obtain the constraints information, *i.e.*, the equation above. The methodology is shown in the figure below.

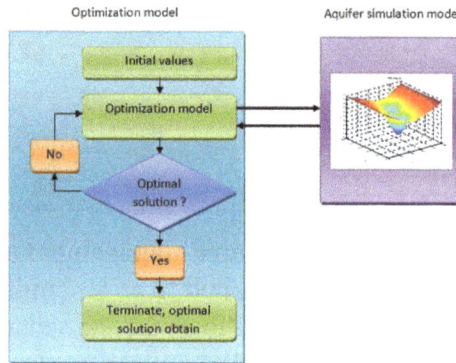

Simulation optimization methodology

The model can be solved using an optimization algorithm.

Considering the aquifer and simulation parameters shown in Table b, and also considering $h_{min} = 99.00m$, the solution of the optimization model is shown in the figure below. The problem is solved for four time steps.

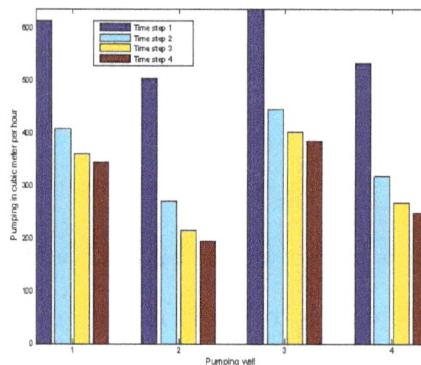

Optimal pumping patterns

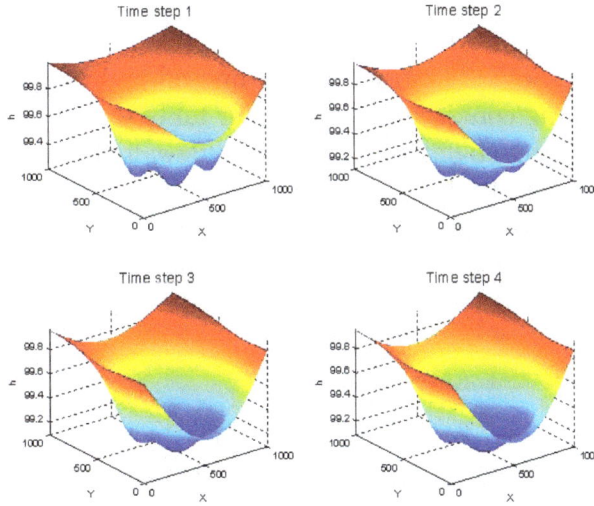

Head distribution in each time step

Optimization Solver in Excel

Various optimization methods are available. Depending on the nature of the problem, one can choose a suitable method to solve the optimization model. The detail discussion about the selection of suitable method is beyond the scope of this lesson note. However, with minimum knowledge of optimization techniques, one can use the optimization solver available in Microsoft excel for solving the groundwater management problem. In this lesson, we will solve an optimization problem using excel solver.

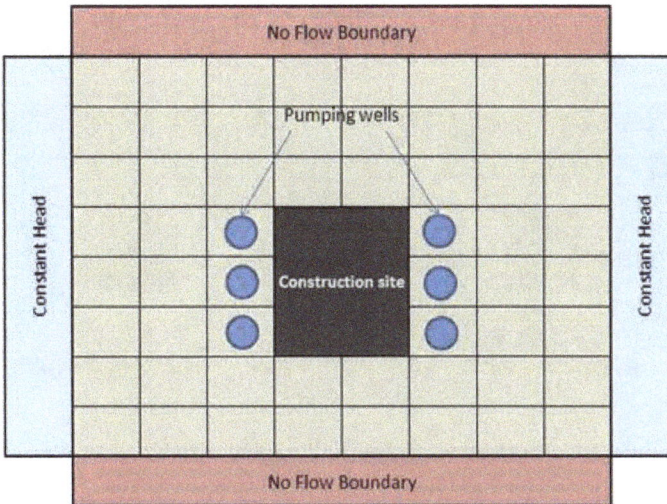

Schematic diagram of a dewatering problem

Consider a two-dimensional unconfined aquifer shown in the figure above. The left and right hand sides of the aquifer have constant head. The other two sides of the aquifer have no flow boundary. Some construction is proposed in the area shown as construction site. To facilitate the construction, the water table has to bring down to certain desirable level. Thus objective of the problem is to dewater the area. This can be done by pumping water from the wells near to the area. In this problem, six wells have been consider for dewatering the area. The area can be dewatered by pumping water from these wells. Now this is an optimization model, as there are numerous combinations of pumping rates from these six wells. This can be solved using an optimization technique.

Formulation of Optimization Model

This is a problem of an unconfined aquifer. As such the flow equation for the unconfined aquifer has to be considered here. The two-dimensional flow equation for homogeneous isotropic unconfined aquifer can be written as,

$$\frac{\partial^2 h^2}{\partial x^2} + \frac{\partial^2 h^2}{\partial y^2} + \frac{2N(xy)}{K} = 0$$

Considering d = h², we have

$$\frac{\partial^2 d}{\partial x^2} + \frac{\partial^2 d}{\partial y^2} + \frac{2N(x,y)}{K} = 0$$

The finite difference form of the equation for any arbitrary location can be written as,

$$\frac{d_{i-1,j} 2d_{i,j} + d_{i+1,j}}{(\Delta x)^2} + \frac{d_{i,j-1} - 2d_{i,j} + d_{i,j-1}}{(\Delta y)^2} + \frac{2}{K} N_{i,j} = 0$$

$$g_{i,j} = \frac{1}{(\Delta x)^2} \left(d_{i-1,j} - 2d_{i,j} + d_{i+1,j} \right) + \frac{1}{(\Delta y)^2} \left(d_{i,j-1} - 2d_{i,j} + d_{i,j-1} \right) + \frac{2}{K} N_{i,j} = 0$$

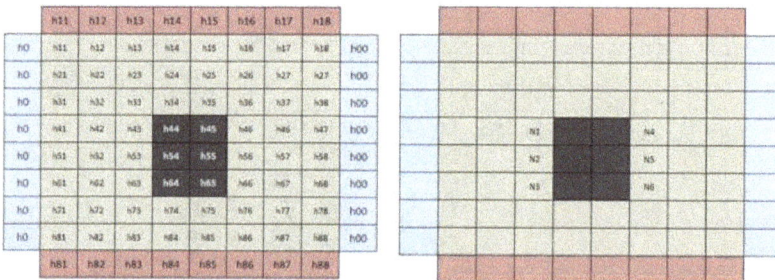

(a) Head variables (b) Pumping variables

Consider the Fig. (a). h_o and h_{oo} is the constant head at left and right boundary of the aquifer. The no flow boundary at top and bottom of the aquifer has been implemented by putting equal heads at the boundary cells as shown in the Fig. (a). Fig. (b) shows the pumping variables. The optimization model can be formulated as,

Minimize $\sum_{k=1}^{6} |N_k|$

Subjectice to

$g_{i,j} = 0 \ for \ i = 1,2,\ldots,8; \ j = 1,2,\ldots,8$

$h_{44} = \sqrt{d_{44}} \leq h_{max}$

$h_{45} = \sqrt{d_{45}} \leq h_{max}$

$h_{54} = \sqrt{d_{54}} \leq h_{max}$

$h_{55} = \sqrt{d_{55}} \leq h_{max}$

$h_{64} = \sqrt{d_{64}} \leq h_{max}$

$h_{65} = \sqrt{d_{65}} \leq h_{max}$

$h_{i,j} \geq 0$

$N_k \geq for \ k = 1,2,\ldots,6$

The optimization problem can be solved using any optimization technique. In this lesson note, the problem has been solved using excel solver available in Microsoft Office. The step by step procedure to solve the problem is given below.

Step 1: Consider the following data

- $h_o = 100m \ and \ h_{oo} = 100m$

- $K = 24, /day$

Step 2: Define the aquifer and aquifer boundary as shown in the figure below. Here left and right boundary has constant head of 100 m. Thus the d will be 10000. The initial trial value of d is also considered as 10000 for the whole aquifer.

Aquifer with initial trial values and boundary values

Step 3: Define actual head value which is equal to square root of d. Figure below shows the h and d values together. The cell under the head values is nothing but the square root of the corresponding cell under the d values.

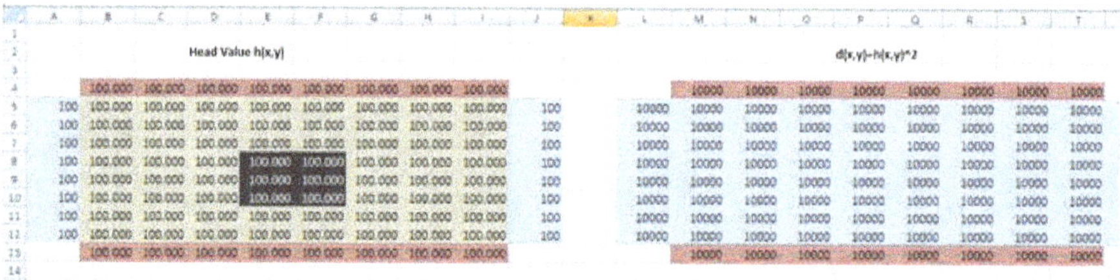

Aquifer with h and d values

Step 4: Now define pumping values as shown in the figure below. Here only six cells have pumping values and rest cell have zero pumping values. The values in the cell will be negative in sign. For recharge the values will be positive. The objective function value is the summation of all the pumping values.

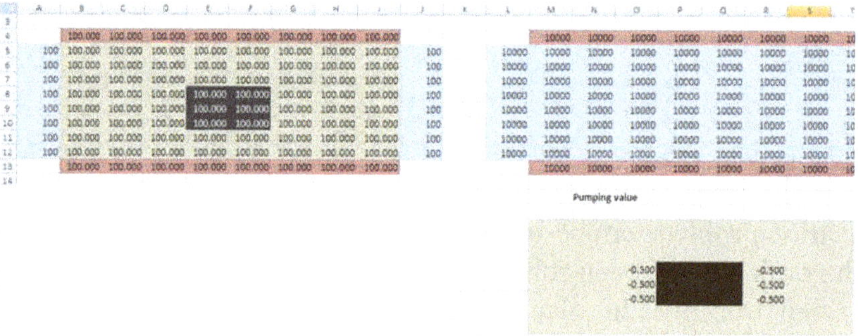

Defined pumping value

Step 5: Now calculate the RHS of the equation for each cell as shown in the figure below. The value of the RHS of the governing equation will be zero in most of the cell but it will not be zero for the pumping cell.

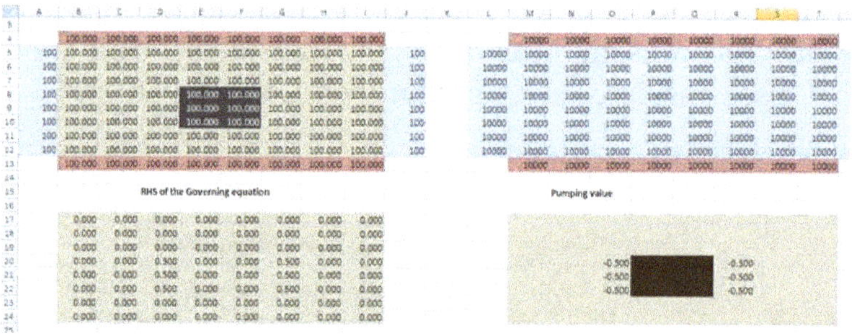

Calculate RHS of the governing equation

Step 6: Now open excel solver by clicking the solver icon available in data menu. Define the cell as shown in the figure below.

Target, Changing and constraint cells of the solver

Step 7: Run the solver by clicking the solve button. The solver will solve the problem and give the optimum solution. The optimum objective function value of the problem is 3.18.

Solution obtained by solver

Step 8: Draw the head contour map (figure below). The head value at the construction site is less than 97 as expected.

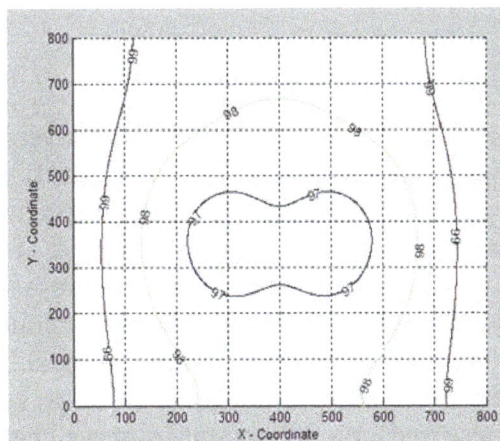

Head contour map

Groundwater Inverse Modeling Using Excel

Inverse modeling technique is generally used to solve parameter estimation problem. For example, one can estimate the aquifer parameters, *i.e.* hydraulic conductivity and storativity of an aquifer using inverse modeling technique. Similarly, inverse modeling technique can also be used for determining illegal pumping from an aquifer. This is called inverse model as we predict the inputs (aquifer parameters, pumping values, *etc.*) from the known outputs (head distribution). As we know that groundwater has been depleted in many parts of the world in last few years because of rapid urbanization. As such there is a need for controlling aquifer abstraction. The uncontrolled abstraction of groundwater may even more dangerous when problems like pollution and saline intrusion are associated with the aquifer, as illegal pumping may exaggerated the problem of groundwater contamination. Therefore, it is the legal duty of the local administration to check the illegal pumping from the aquifer. The illegal pumping of an aquifer can be estimated using groundwater flow inverse model.

Formulation and Solution of Groundwater Inverse Model

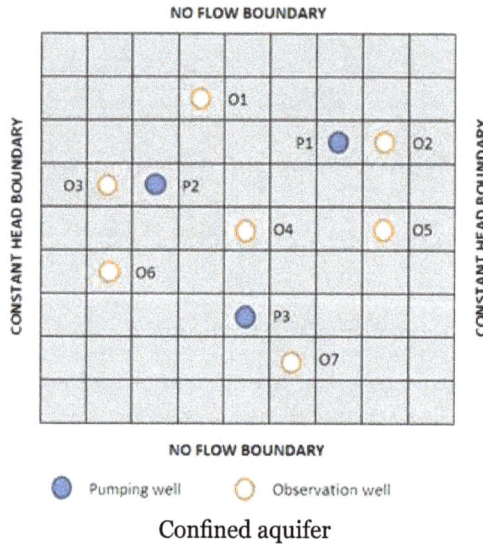

Confined aquifer

In this section, we will formulate and solve a groundwater inverse problem. The inverse model estimates pumping rate of the installed wells in a confined aquifer from known hydraulic head values which are monitored at the observation wells. Figure above shows a hypothetical confined aquifer of horizontal extend of 1000mx1000m. There are three pumping wells which are marked as P1 to P3. Pumping from these wells may be restricted by the local authority under some managerial policies. As such, there is a need to estimate these pumping values by monitoring hydraulic head at some observation locations. There are seven observation wells available in the aquifer, which are marked as O1 to O7. At these wells, the steady state observed head values have been recorded. Random noise is added to the observed head values for simulating observation errors. Table below shows the observed head values at observation wells. The initial head values are generated randomly.

Observation head

Observation Well	O1	O2	O3	O4	O5	O6	O7
Observed Head	92.84	94.08	93.48	92.03	95.67	95.43	93.80

The north and south boundaries of the aquifer are no flow boundary and other two sides, *i.e.* east and west sides are constant head boundary. The aquifer is homogeneous and isotropic with transmissivity (*T*) of 300 m² /day. The objective of the problem is to estimate the rate of pumping from the pumping wells from known value of steady state hydraulic head at the observation wells. As the aquifer is confined, we need to use the flow equation of confined aquifer. We have already discussed the 2-D steady flow equation for homogeneous isotropic confined aquifer which may be written as

$$TV^2\varphi + N(x, y) = 0$$

Where, *T* is the transmissivity of the aquifer (m^2/day), and V^2 is the Laplace operator, φ is the hydraulic head *(m)*, *N* is the pumping or recharge value $(m^3/day/m^2)$. For using finite difference method, the aquifer has been discretized with grid size of 111.1m×111.1m, i.e. $\Delta x = \Delta y = 111.1$m. The finite difference approximation of the steady 2-D flow equation at any cell (i, j) may be written as,

$$A\varphi_{i+1,j} + A\varphi_{i-1,j} + B\varphi_{i,j+1} + B\varphi_{i,j-1} - (2A + 2B)\varphi_{i,j} + N(i, j) = 0$$

Where $A = \left(T / (\Delta x)^2\right), B = \left(T / (\Delta y)^2\right), \Delta x \, (m)$ is the grid size in *x* direction and Δy(m)

is the grid size in *y* direction. The groundwater flow can be simulated using optimization based simulation approach. The simulation model solves the set of discretized equations written at each grid centre of the discretized aquifer. The optimization based simulation model can be written as,

Minimize $f = \sum_{i=1}^{I} \sum_{j=1}^{J} e_{i,j}^2$

Subject to $e_{i,j} = A\varphi_{i+1,j} + A\varphi_{i-1,j} + B\varphi_{i,j+1} + B\varphi_{i,j-1} - (2A + 2B)B\varphi_{i,j} + N(i, j) = 0$

$\varphi_{i,j} \geq 0$

Where i =1.2.3....., *I* and j =1.2.3....., *J*, *I* and *J* are the total number of columns and rows of the discretized aquifer.

The inverse problem for estimating pumping value can be formulated using optimization approach. The optimization model minimizes the difference between simulated and observed hydraulic head values of given observation locations. The finite difference approximation of the flow equation is added as embedded constraints to the optimization model along with non negative restriction of the decision variables. The optimization formulation may be written as,

$$\text{Minimize } f = \sum_{i=1}^{I} \sum_{j=1}^{J} e_{i,j}^2 + \sum_{k=1}^{K} \left(\varphi_k^o - \varphi_k^c \right), \quad k = 1, 2, 3, \ldots, K$$

$$\text{Subject to } e_{i,j} = 0 \qquad i = 1, 2, 3, \ldots, I \text{ and } j = 1, 2, 3, \ldots, J$$

$$\varphi_k^c \geq 0 \qquad k = 1, 2, 3, \ldots, K$$

$$\varphi_{i,j} \geq 0 \qquad i = 1, 2, 3, \ldots, I \text{ and } j = 1, 2, 3, \ldots, J$$

Where, φ_k^o and φ_k^c are the observed and computed hydraulic head values at k^{th} observation location, $e_{i,j}$ is the right hand side of the finite difference approximation of flow equation, K is the total number of the observation locations. The solution of the optimization model will give the unknown pumping rates of the pumping wells of the aquifer. Apart from the unknown pumping rates, additionally we can also obtain the head distribution of the aquifer.

The optimization problem can be solved using any optimization algorithm solver. However, here the optimization model is solved using excel solver. The initial value of head are generated randomly between 90m and 100m. The solutions of the optimization model are shown in table. It may be observed from the table that the actual pumping and estimated pumping values are almost same. For example, the true pumping from P1 well is 4000 m^3/day and the estimated pumping is 4040 m^3/day. In case of pumping well P2, the true and estimated pumping are 5000 m³/day and 5053 m³/day.

Table: Observed and Predicted Pumping value with relative error

Pumping Wells	Actual Pumping (m3/day)	Predicted Pumping (m3/day)	Relative Error (%)
P1	4000	4040	1.00
P2	5000	5053	1.06
P3	3000	3044	1.47

Similarly, for the third pumping well also the difference between true and estimated pumping rate is very small and in acceptable range. The relative errors between true and estimated pumping values are also very small. Maximum relative error of 1.47% is encountered at pumping well. As discussed earlier, the embedded approach also gives distribution of hydraulic head along with the optimal solutions. Figure below shows the contour map of hydraulic head at optimal solutions.

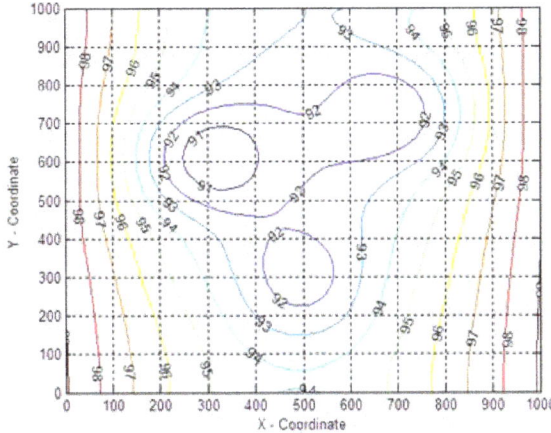

Contour map showing head distribution at optimal solution

Safe Yield

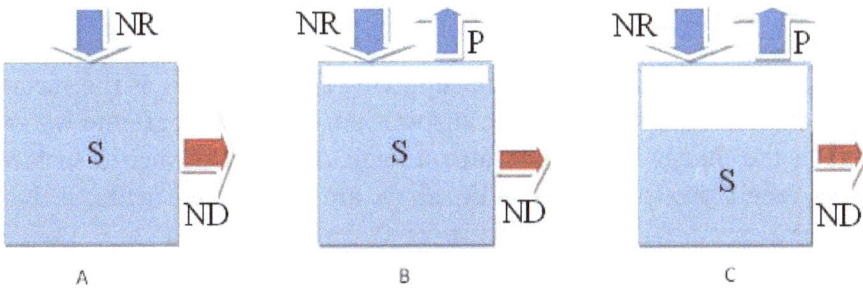

Effect of withdrawn on groundwater balance

The abstraction of water from underground aquifer at a greater rate than it is being re-charged leads to the lowering of water table and upsets the equilibrium between dis-charge and recharge. If we compare groundwater with surface water, surface water is re-newable in nature. Generally renew within few days to weeks. On the other hand, through groundwater is also renewable in nature, it takes longer time to renew, sometime up to decades or centuries. The concept of safe yield has been used to express the quantity of water that can be withdrawn from the ground without damaging an aquifer as a water source. The damage includes (i) reduction in groundwater withdrawn, (ii) reduction in ecological base flow, (iii) land subsidence, (iv) depletion of groundwater reservoir, *etc* . The definition of safe yield can also be written as "The maximum quantity of water that can be guaranteed during a critical dry period is known as the safe yield or firm yield." The definition given by Todd (1959) is "Safe yield is the basin draft on a groundwater supply which can be continued indefinitely without harming the supply or basin landowners".

Figure above shows the effect of withdrawn of water on groundwater balance. The Fig. (A) shows the natural condition of an aquifer. In this case, the amount of water recharge naturally is equal to the amount of water drain out from the aquifer. The aquifer is naturally at steady condition. This is also known as pristine groundwater system. The Fig. (B) shows the condition of the aquifer under stress condition. In this case, the pumping is less and thus there is balance among the amount of water recharge, the natural drainage and the amount of water withdrawn. The aquifer is in equilibrium condition and this system is known as developed groundwater system. The Fig. (C) shows the condition when water withdrawn from an aquifer is much higher than the natural recharge. In this case, the storage of the aquifer will be depleted with time and the aquifer may not be in a position to yield any water in future. In this case, the draft is more than the safe yield of the aquifer which is called as overdraft. This system is known as depleted groundwater system. This can give rise to pollution or cause serious problems due to severely increasing pumping lift. Indeed, this eventually may lead to the exhaustion of a well.

Estimation of safe yield is a complex problem and should consider the climatic, geological and hydrological conditions. Due to the variation of climatic and hydrologic conditions with time, the safe yield is likely to vary appreciably with time. The safe yield, G, often is expressed as follows:

$$G = P - Q_s - E_T + Q_g - \Delta S_g - \Delta S_s$$

Where, P is the precipitation on the area supplying the aquifer, Q_s is the surface stream flow over the same area, E_T is the evapotranspiration, Q_g is the net ground water inflow to the area, ΔS_g is the change in ground water storage, ΔS_s is the change of surface storage. All the terms, except precipitation of in Eq. above are subjected to artificial change.

Stream-aquifer Interaction

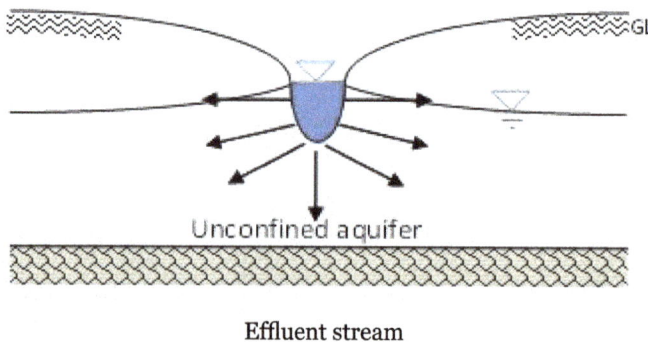

Effluent stream

River, in general contributes water to the adjacent phreatic aquifer. The low water flow in a river is nothing but the contribution of groundwater from the adjacent aquifer.

The low water flow in a river is also known as the base flow of the river. In this case, the elevation of the groundwater table is higher than the elevation of the stream. When groundwater is feeding a river, the stream is known as effluent stream.

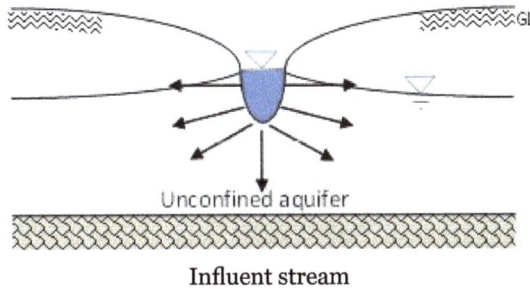

Influent stream

On the other hand when elevation of the groundwater table is lower than the elevation of the stream surface, water will flow from the stream to the aquifer. In this case, stream is feeding the aquifer and the stream is known as influent stream. Figure above shows an influent stream. In some cases, the groundwater table is located below the channel bottom as shown in figure below. In this case also, the stream feeds the aquifer and the stream is known as influent stream.

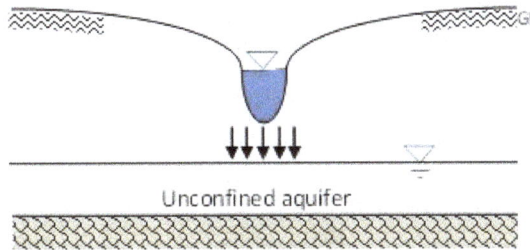

Influent stream where GW table is below the bottom of the channel

Sometime a stream may act as an effluent stream as well as influent stream. Figure below shows such a stream where on one side of the stream, the groundwater table is higher than the stream surface and on the other side of the stream, the elevation of the groundwater table is lower than the stream surface. Therefore, one side of the stream acts as an effluent stream and the other side acts as an influent stream.

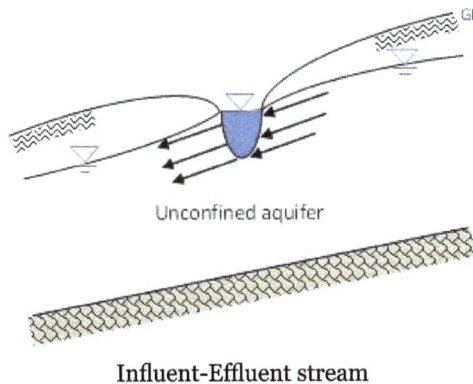

Influent-Effluent stream

The above discussion shows that the interaction of groundwater with stream/river has an important role in management of groundwater resources. Thus river plays an important role in solving groundwater management problem. One primary role of a river is that river acts as a constant head boundary condition in solving the groundwater forecasting problem. River also acts as the source of water which contributes water to the aquifer.

Groundwater Recharge

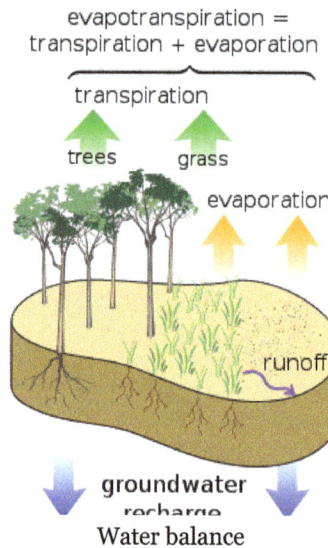

Water balance

Groundwater recharge or deep drainage or deep percolation is a hydrologic process where water moves downward from surface water to groundwater. Recharge is the primary method through which water enters an aquifer. This process usually occurs in the vadose zone below plant roots and is often expressed as a flux to the water table surface. Recharge occurs both naturally (through the water cycle) and through anthropogenic processes.

Processes

Groundwater is recharged naturally by rain and snow melt and to a smaller extent by surface water (rivers and lakes). Recharge may be impeded somewhat by human activities including paving, development, or logging. These activities can result in loss of topsoil resulting in reduced water infiltration, enhanced surface runoff and reduction in recharge. Use of groundwaters, especially for irrigation, may also lower the water tables. Groundwater recharge is an important process for sustainable groundwater management, since the volume-rate abstracted from an aquifer in the long term should be less than or equal to the volume-rate that is recharged.

Recharge can help move excess salts that accumulate in the root zone to deeper soil layers, or into the groundwater system. Tree roots increase water saturation into groundwater reducing water runoff. Flooding temporarily increases river bed permeability by moving clay soils downstream, and this increases aquifer recharge.

Artificial groundwater recharge is becoming increasingly important in India, where over-pumping of groundwater by farmers has led to underground resources becoming depleted. In 2007, on the recommendations of the International Water Management Institute, the Indian government allocated ₹1,800 crore (equivalent to $38 billion or US$590 million in 2016) to fund dug-well recharge projects (a dug-well is a wide, shallow well, often lined with concrete) in 100 districts within seven states where water stored in hard-rock aquifers had been over-exploited. Another environmental issue is the disposal of waste through the water flux such as dairy farms, industrial, and urban runoff.

Wetlands

Wetlands help maintain the level of the water table and exert control on the hydraulic head (O'Brien 1988; Winter 1988). This provides force for groundwater recharge and discharge to other waters as well. The extent of groundwater recharge by a wetland is dependent upon soil, vegetation, site, perimeter to volume ratio, and water table gradient (Carter and Novitzki 1988; Weller 1981). Groundwater recharge occurs through mineral soils found primarily around the edges of wetlands (Verry and Timmons 1982) The soil under most wetlands is relatively impermeable. A high perimeter to volume ratio, such as in small wetlands, means that the surface area through which water can infiltrate into the groundwater is high (Weller 1981). Groundwater recharge is typical in small wetlands such as prairie potholes, which can contribute significantly to recharge of regional groundwater resources (Weller 1981). Researchers have discovered groundwater recharge of up to 20% of wetland volume per season (Weller 1981).

Depression-focused Recharge

If water falls uniformly over a field such that field capacity of the soil is not exceeded, then negligible water percolates to groundwater. If instead water puddles in low-lying areas, the same water volume concentrated over a smaller area may exceed field capacity resulting in water that percolates down to recharge groundwater. The larger the relative contributing runoff area is, the more focused infiltration is. The recurring process of water that falls relatively uniformly over an area, flowing to groundwater selectively under surface depressions is depression focused recharge. Water tables rise under such depressions.

Depression Pressure

Depression focused groundwater recharge can be very important in arid regions. More rain events are capable of contributing to groundwater supply.

Depression focused groundwater recharge also profoundly effects contaminant transport into groundwater. This is of great concern in regions with karst geological formations because water can eventually dissolve tunnels all the way to aquifers, or otherwise disconnected streams. This extreme form of preferential flow, accelerates the transport of contaminants and the erosion of such tunnels. In this way depressions intended to trap runoff water—before it flows to vulnerable water resources—can connect underground over time. Cavitation of surfaces above into the tunnels, results in potholes or caves.

Deeper ponding exerts pressure that forces water into the ground faster. Faster flow dislodges contaminants otherwise adsorbed on soil and carries them along. This can carry pollution directly to the raised water table below and into the groundwater supply. Thus the quality of water collecting in infiltration basins is of special concern.

Pollution

Pollution in stormwater runoff collects in retention basins. Concentrating degradable contaminants can accelerate biodegradation. However, where and when water tables are high this affects appropriate design of detention ponds, retention ponds and rain gardens.

Estimation Methods

Rates of groundwater recharge are difficult to quantify since other related processes, such as evaporation, transpiration (or evapotranspiration) and infiltration processes must first be measured or estimated to determine the balance.

Physical

Physical methods use the principles of soil physics to estimate recharge. The *direct* physical methods are those that attempt to actually measure the volume of water passing below the root zone. *Indirect* physical methods rely on the measurement or estimation of soil physical parameters, which along with soil physical principles, can be used to estimate the potential or actual recharge. After months without rain the level of the rivers under humid climate is low and represents solely drained groundwater. Thus, the recharge can be calculated from this base flow if the catchment area is known.

Chemical

Chemical methods use the presence of relatively inert water-soluble substances, such as an isotopic tracer or chloride, moving through the soil, as deep drainage occurs.

Numerical Models

Recharge can be estimated using numerical methods, using such codes as Hydrologic

Evaluation of Landfill Performance, UNSAT-H, SHAW, WEAP, and MIKE SHE. The 1D-program HYDRUS1D is available online. The codes generally use climate and soil data to arrive at a recharge estimate and use the Richards equation in some form to model groundwater flow in the vadose zone.

Adverse Factors

- Drainage
- Impervious surfaces
- Soil compaction
- Groundwater pollution

Groundwater Remediation

Groundwater remediation is the process that is used to treat polluted groundwater by removing the pollutants or converting them into harmless products. Groundwater is water present below the ground surface that saturates the pore space in the subsurface. Globally, between 25 per cent and 40 per cent of the world's drinking water is drawn from boreholes and dug wells. Groundwater is also used by farmers to irrigate crops and by industries to produce everyday goods. Most groundwater is clean, but groundwater can become polluted, or contaminated as a result of human activities or as a result of natural conditions.

The many and diverse activities of humans produce innumerable waste materials and by-products. Historically, the disposal of such waste have not been subject to many regulatory controls. Consequently, waste materials have often been disposed of or stored on land surfaces where they percolate into the underlying groundwater. As a result, the contaminated groundwater is unsuitable for use.

Current practices can still impact groundwater, such as the over application of fertilizer or pesticides, spills from industrial operations, infiltration from urban runoff, and leaking from landfills. Using contaminated groundwater causes hazards to public health through poisoning or the spread of disease, and the practice of groundwater remediation has been developed to address these issues. Contaminants found in groundwater cover a broad range of physical, inorganic chemical, organic chemical, bacteriological, and radioactive parameters. Pollutants and contaminants can be removed from groundwater by applying various techniques, thereby bringing the water to a standard that is commensurate with various intended uses.

Techniques

Ground water remediation techniques span biological, chemical, and physical treatment technologies. Most ground water treatment techniques utilize a combination of technologies. Some of the biological treatment techniques include bioaugmentation, bioventing, biosparging, bioslurping, and phytoremediation. Some chemical treatment techniques include ozone and oxygen gas injection, chemical precipitation, membrane separation, ion exchange, carbon absorption, aqueous chemical oxidation, and surfactant enhanced recovery. Some chemical techniques may be implemented using nanomaterials. Physical treatment techniques include, but are not limited to, pump and treat, air sparging, and dual phase extraction.

Biological Treatment Technologies

Bioaugmentation

If a treatability study shows no degradation (or an extended lab period before significant degradation is achieved) in contamination contained in the groundwater, then inoculation with strains known to be capable of degrading the contaminants may be helpful. This process increases the reactive enzyme concentration within the bioremediation system and subsequently may increase contaminant degradation rates over the nonaugmented rates, at least initially after inoculation.

Bioventing

Bioventing is an in situ remediation technology that uses microorganisms to biodegrade organic constituents in the groundwater system. Bioventing enhances the activity of indigenous bacteria and archaea and stimulates the natural in situ biodegradation of hydrocarbons by inducing air or oxygen flow into the unsaturated zone and, if necessary, by adding nutrients. During bioventing, oxygen may be supplied through direct air injection into residual contamination in soil. Bioventing primarily assists in the degradation of adsorbed fuel residuals, but also assists in the degradation of volatile organic compounds (VOCs) as vapors move slowly through biologically active soil.

Biosparging

Biosparging is an in situ remediation technology that uses indigenous microorganisms to biodegrade organic constituents in the saturated zone. In biosparging, air (or oxygen) and nutrients (if needed) are injected into the saturated zone to increase the biological activity of the indigenous microorganisms. Biosparging can be used to reduce concentrations of petroleum constituents that are dissolved in groundwater, adsorbed to soil below the water table, and within the capillary fringe.

Bioslurping

Bioslurping combines elements of bioventing and vacuum-enhanced pumping of free-product that is lighter than water (light non-aqueous phase liquid or LNAPL) to recover free-product from the groundwater and soil, and to bioremediate soils. The bioslurper system uses a "slurp" tube that extends into the free-product layer. Much like a straw in a glass draws liquid, the pump draws liquid (including free-product) and soil gas up the tube in the same process stream. Pumping lifts LNAPLs, such as oil, off the top of the water table and from the capillary fringe (i.e., an area just above the saturated zone, where water is held in place by capillary forces). The LNAPL is brought to the surface, where it is separated from water and air. The biological processes in the term "bioslurping" refer to aerobic biological degradation of the hydrocarbons when air is introduced into the unsaturated zone.

Phytoremediation

In the phytoremediation process certain plants and trees are planted, whose roots absorb contaminants from ground water over time, and are harvested and destroyed. This process can be carried out in areas where the roots can tap the ground water. Few examples of plants that are used in this process are Chinese Ladder fern Pteris vittata, also known as the brake fern, is a highly efficient accumulator of arsenic. Genetically altered cottonwood trees are good absorbers of mercury and transgenic Indian mustard plants soak up selenium well.

Permeable Reactive Barriers

Certain types of permeable reactive barriers utilize biological organisms in order to remediate groundwater.

Chemical Treatment Technologies

Chemical Precipitation

Chemical precipitation is commonly used in wastewater treatment to remove hardness and heavy metals. In general, the process involves addition of agent to an aqueous waste stream in a stirred reaction vessel, either batchwise or with steady flow. Most metals can be converted to insoluble compounds by chemical reactions between the agent and the dissolved metal ions. The insoluble compounds (precipitates) are removed by settling and/or filtering.

Ion Exchange

Ion exchange for ground water remediation is virtually always carried out by passing the water downward under pressure through a fixed bed of granular medium (either cation exchange media and anion exchange media) or spherical beads. Cations are displaced by certain cations from the solutions and ions are displaced by certain anions from the solution. Ion exchange media most often used for remediation are zeolites (both natural and synthetic) and synthetic resins.

Carbon Absorption

The most common activated carbon used for remediation is derived from bituminous coal. Activated carbon absorbs volatile organic compounds from ground water by chemically binding them to the carbon atoms.

Chemical Oxidation

In this process, called In Situ Chemical Oxidation or ISCO, chemical oxidants are delivered in the subsurface to destroy (converted to water and carbon dioxide or to nontoxic substances) the organics molecules. The oxidants are introduced as either liquids or gasses. Oxidants include air or oxygen, ozone, and certain liquid chemicals such as hydrogen peroxide, permanganate and persulfate. Ozone and oxygen gas can be generated on site from air and electricity and directly injected into soil and groundwater contamination. The process has the potential to oxidize and/or enhance naturally occurring aerobic degradation. Chemical oxidation has proven to be an effective technique for dense non-aqueous phase liquid or DNAPL when it is present.

Surfactant Enhanced Recovery

Surfactant enhanced recovery increases the mobility and solubility of the contaminants absorbed to the saturated soil matrix or present as dense non-aqueous phase liquid. Surfactant-enhanced recovery injects surfactants (surface-active agents that are primary ingredient in soap and detergent) into contaminated groundwater. A typical system uses an extraction pump to remove groundwater downstream from the injection point. The extracted groundwater is treated aboveground to separate the injected surfactants from the contaminants and groundwater. Once the surfactants have separated from the groundwater they are re-used. The surfactants used are non-toxic, food-grade, and biodegradable. Surfactant enhanced recovery is used most often when the groundwater is contaminated by dense non-aqueous phase liquids (DNAPLs). These dense compounds, such as trichloroethylene (TCE), sink in groundwater because they have a higher density than water. They then act as a continuous source for contaminant plumes that can stretch for miles within an aquifer. These compounds may biodegrade very slowly. They are commonly found in the vicinity of the original spill or leak where capillary forces have trapped them.

Permeable Reactive Barriers

Some permeable reactive barriers utilize chemical processes to achieve groundwater remediation.

Physical Treatment Technologies

Pump and Treat

Pump and treat is one of the most widely used ground water remediation technologies.

In this process ground water is pumped to the surface and is coupled with either biological or chemical treatments to remove the impurities.

Air Sparging

Air sparging is the process of blowing air directly into the ground water. As the bubbles rise, the contaminants are removed from the groundwater by physical contact with the air (i.e., stripping) and are carried up into the unsaturated zone (i.e., soil). As the contaminants move into the soil, a soil vapor extraction system is usually used to remove vapors.

Dual Phase Vacuum Extraction

Dual-phase vacuum extraction (DPVE), also known as multi-phase extraction, is a technology that uses a high-vacuum system to remove both contaminated groundwater and soil vapor. In DPVE systems, a high-vacuum extraction well is installed with its screened section in the zone of contaminated soils and groundwater. Fluid/vapor extraction systems depress the water table and water flows faster to the extraction well. DPVE removes contaminants from above and below the water table. As the water table around the well is lowered from pumping, unsaturated soil is exposed. This area, called the capillary fringe, is often highly contaminated, as it holds undissolved chemicals, chemicals that are lighter than water, and vapors that have escaped from the dissolved groundwater below. Contaminants in the newly exposed zone can be removed by vapor extraction. Once above ground, the extracted vapors and liquid-phase organics and groundwater are separated and treated. Use of dual-phase vacuum extraction with these technologies can shorten the cleanup time at a site, because the capillary fringe is often the most contaminated area.

Monitoring-Well Oil Skimming

Monitoring-wells are often drilled for the purpose of collecting ground water samples for analysis. These wells, which are usually six inches or fewer in diameter, can also be used to remove hydrocarbons from the contaminant plume within a groundwater aquifer by using a belt style oil skimmer. Belt oil skimmers, which are simple in design, are commonly used to remove oil and other floating hydrocarbon contaminants from industrial water systems.

A monitoring-well oil skimmer remediates various oils, ranging from light fuel oils such as petrol, light diesel or kerosene to heavy products such as No. 6 oil, creosote and coal tar. It consists of a continuously moving belt that runs on a pulley system driven by an electric motor. The belt material has a strong affinity for hydrocarbon liquids and for shedding water. The belt, which can have a vertical drop of 100+ feet, is lowered into the monitoring well past the LNAPL/water interface. As the belt moves through this interface it picks up liquid hydrocarbon contaminant, which is removed and collected at ground level as the belt passes through a wiper mechanism. To the extent that DNAPL hydrocarbons settle at the bottom of a monitoring well, and the lower pulley of the belt skimmer reaches them, these contaminants can also be removed by a monitoring-well oil skimmer.

Typically, belt skimmers remove very little water with the contaminant, so simple weir type separators can be used to collect any remaining hydrocarbon liquid, which often makes the water suitable for its return to the aquifer. Because the small electric motor uses little electricity, it can be powered from solar panels or a wind turbine, making the system self-sufficient and eliminating the cost of running electricity to a remote location.

Watertable Control

Watertable control is the practice of controlling the height of the water table by drainage. Its main applications are in agricultural land (to improve the crop yield using agricultural drainage systems) and in cities to manage the extensive underground infrastructure that includes the foundations of large buildings, underground transit systems, and extensive utilities (water supply networks, sewerage, storm drains, and underground electrical grids).

Description and Definitions

Subsurface land drainage aims at controlling the water table of the groundwater in originally waterlogged land at a depth acceptable for the purpose for which the land is used. The depth of the water table with drainage is *greater* than without.

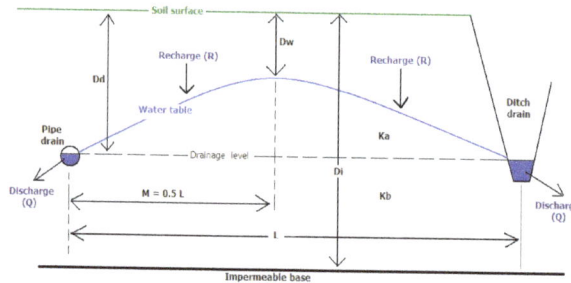

Geometry subsurface drainage system by pipes or ditches
D = depth   K = hydraulic conductivity   L = Drain spacing

Drainage parameters in watertable control

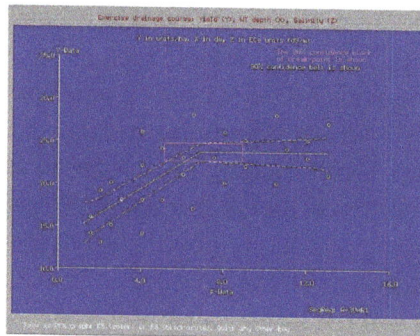

Crop yield and depth of water table (X in dm)

Purpose

In agricultural land drainage, the purpose of water table control is to establish a depth of the water table that does no longer interfere negatively with the necessary farm operations and crop yields (made with the SegReg model).

In addition, land drainage can help with soil salinity control.

The soil's hydraulic conductivity plays an important role in drainage design.

The development of *agricultural drainage criteria* is required to give the designer and manager of the drainage system a target to achieve in terms of maintenance of an optimum depth of the water table.

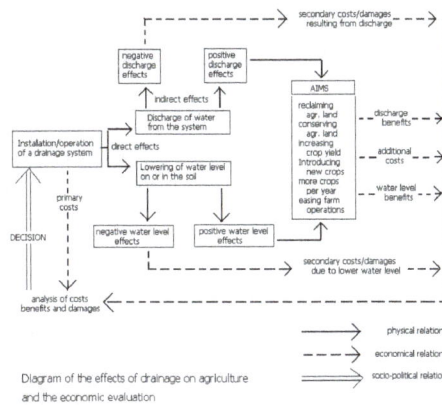

Diagram of the effects of drainage on agriculture
and the economic evaluation

Positive and negative effects of land drainage

Optimization

Optimization of the depth of the water table is related the benefits and costs of the drainage system. The shallower the permissible depth of the water table, the lower the cost of the drainage system to be installed to achieve this depth. However, the lowering of the originally too shallow depth by land drainage entails *side effects*. These have also to be taken into account, including the costs of mitigation of negative side effects.

Drain depth (D_d, m), soil salinity (C_r, dS/m),
field irrigation efficiency of the group A crops (FaA, -),
field irrigation sufficiency of the group A crops (JsA, -),
seasonal average depth of the water table (D_w, m), and
quantity of drainage water (G_d, mm per season).

Drain Depth	1st season (summer)				
D_d	C_r	FaA	JsA	D_w	G_d
0.6	2.7	0.84	0.99	0.37	105
0.8	2.5	0.83	0.98	0.55	112
1.0	2.4	0.82	0.97	0.74	117
1.2	2.2	0.81	0.96	0.93	122
1.4	2.1	0.80	0.95	1.12	127
	2nd season (winter)				
0.6	2.8	0.86	0.97	0.55	31
0.8	2.7	0.84	0.95	0.74	37
1.0	2.5	0.82	0.93	0.94	45
1.2	2.3	0.81	0.92	1.12	54
1.4	2.2	0.80	0.91	1.31	57

Example of effects of drain depth

The *optimization* of drainage design and the development of *drainage criteria* are discussed in the article on drainage research.

Figure above shows an example of the effect of drain depth on soil salinity and various irrigation/drainage parameters as simulated by the SaltMod program.

History

Historically, agricultural land drainage started with the digging of relatively shallow open ditches that received both runoff from the land surface and outflow of groundwater. Hence the ditches had a surface as well as a subsurface drainage function.

By the end of the 19th century and early in the 20th century it was felt that the ditches were a hindrance for the farm operations and the ditches were replaced by buried lines of clay pipes (tiles), each tile about 30 cm long. Hence the term "tile drainage".

Since 1960, one started using long, flexible, corrugated plastic (PVC or PE) pipes that could be installed efficiently in one go by trenching machines. The pipes could be pre-wrapped with an envelope material, like synthetic fibre and geotextile, that would prevent the entry of soil particles into the drains.

Thus, land drainage became a powerful industry. At the same time agriculture was steering towards maximum productivity, so that the installation of drainage systems came in full swing.

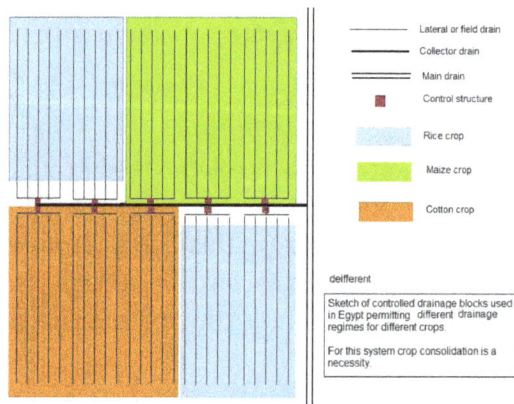

Controlled drainage

Environment

As a result of large scale developments, many modern drainage projects were *over-designed*, while the negative environmental side effects were ignored. In circles with environmental concern, the profession of land drainage got a poor reputation, sometimes justly so, sometimes unjustified, notably when land drainage was confused with the more encompassing activity of wetland reclamation. Nowadays, in some countries, the hardliner trend is reversed. Further, *checked* or *controlled* drainage systems were introduced, as shown in Figure above.

Drainage Design

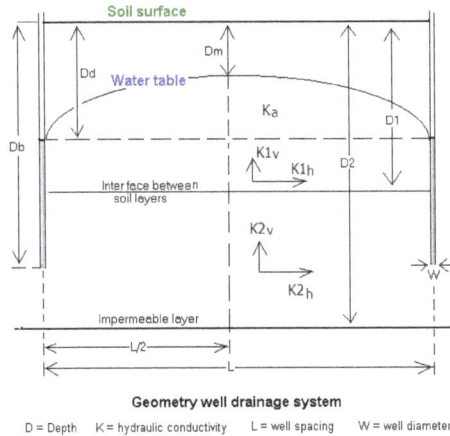

Geometry of a well drainage system

The design of subsurface drainage systems in terms layout, depth and spacing of the drains is often done using subsurface drainage equations with parameters like drain depth, depth of the water table, soil depth, hydraulic conductivity of the soil and drain discharge. The drain discharge is found from an agricultural water balance. The computations can be done using computer models like EnDrain.

Drainage by Wells

Subsurface drainage of groundwater can also be accomplished by pumped wells (*vertical* drainage, in contrast to *horizontal* drainage). Drainage wells have been used extensively in the Salinity Control and Reclamation Program (SCARP) in the Indus valley of Pakistan. Although the experiences were not overly successful, the feasibility of this technique in areas with deep and permeable aquifers is not to be discarded. The well spacings in these areas can be so wide (more than 1000m) that the installation of *vertical* drainage systems could be relatively cheap compared to *horizontal* subsurface drainage (drainage by pipes, ditches, trenches, at a spacing of 100m or less). For the design of a well field for control of the water table, the WellDrain model may be helpful.

References

- Allison, G.B.; Hughes, M.W. (1978). "The use of environmental chloride and tritium to estimate total recharge to an unconfined aquifer". Australian Journal of Soil Research. 16 (2): 181–195. doi:10.1071/SR9780181

- "Urban Trees Enhance Water Infiltration". Fisher, Madeline. The American Society of Agronomy. November 17, 2008. Retrieved October 31, 2012

- Agricultural Drainage Criteria, Chapter 17 in: H.P.Ritzema (2006), Drainage Principles and Applications, Publication 16, International Institute for Land Reclamation and Improvement (ILRI), Wageningen, The Netherlands. ISBN 90-70754-33-9

- Allison, G.B.; Gee, G.W.; Tyler, S.W. (1994). "Vadose-zone techniques for estimating groundwater recharge in arid and semiarid regions". Soil Science Society of America Journal. 58: 6–14

- "Major floods recharge aquifers". University of New South Wales Science. January 24, 2011. Retrieved October 31, 2012

5

Groundwater Contamination: Analysis and Measurement

Groundwater can be contaminated through various sources such as radioactive waste disposal sites, septic tanks and cesspools, mine wastes, landfill leaching, animal burials, pesticide and fertilizers applied to crop field, etc. The contaminating solute in groundwater can spread through dispersion and advection. The aspects elucidated in this section are of vital importance, and provide a better understanding of pollutant flow.

Groundwater Contamination

The groundwater may be contaminated from various sources. Some of the sources are natural and some of them are manmade. Some of the sources of groundwater contamination are:

- Septic tanks and cesspools

- Injection wells of hazardous wastes, agricultural and urban runoff, municipal sewage, etc

- Landfill leaching

- Mine wastes

- Animal burials

- Radioactive waste disposal sites

- Pesticide and fertilizer applied to crop field

- Saltwater intrusion in coastal aquifer

- Leaching of natural minerals like fluoride, arsenic, iron, etc.

The solute present in groundwater aquifer can move from one place to another place mainly by the processes of diffusion and advection. The mixing of solute also takes place by the process of mechanical dispersion. The transportation of pollutants from one place to another place will contaminate the aquifer and the aquifer may become unusable for domestic, industrial, irrigational,etc. uses. As such modeling of the transport processes is essential to mitigate the aquifer contamination problem. In the following sections, we will discuss the processes of mixing of solute in groundwater and also their mathematical modeling.

Molecular Diffusion

Solute in water moves from the areas of higher concentration to the areas of lower concentration. This phenomenon is known as molecular diffusion. This phenomenon can be observed by a simple experiment as shown in the figure below. The blue color of the Fig. (a) shows the solute mixed water and the white color shows the pure water. These two fluids are separated by an impermeable thin separator. At this stage, the distance verses concentration curve will be a step curve. If we remove the impermeable barrier instantaneously, these two fluids will mix each other by the process of diffusion. Fig. (b) shows the mixing of fluid after some time. The distance verses relative concentration curve will now take a shape of S (Fig. d). The diffusion phenomenon can also be observed by putting a drop of ink in a glass of water. The ink will mix with the water as time elapses .

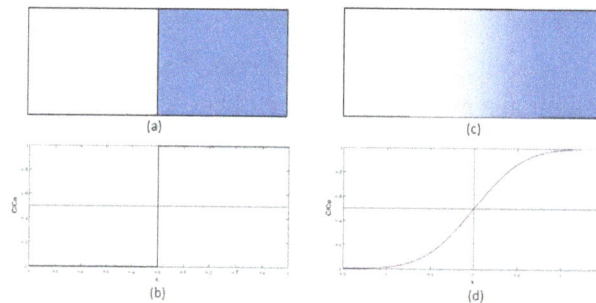

The diffusion process, (a) At initial state, (b) The distance verses relative concentration plot at initial state, (c) At time " t " from the start of the diffusion process, and (d) The distance verses relative concentration plot at time " t "

Fick's First Law

The first Fick's law suggested that the mass of fluid diffusing is proportional to the concentration gradient. As per the law,

$$F \propto -\frac{\partial C}{\partial x}$$

$$F = -D_d \frac{\partial C}{\partial x}$$

Where, F is the mass flux of solute per unit area per unit time, D_d is the diffusion coefficient (L^2/T), C is the solute concentration (M/L^3), $\partial c/\partial x$ is the concentration gradient ($M/L^3/L$). The negative sign indicates that the movement of solute is from higher concentration to lower concentration. This law is similar to Darcy's law where flow is take place from the higher head to lower head.

Fick's Second Law

Consider the control volume shown in the figure below:

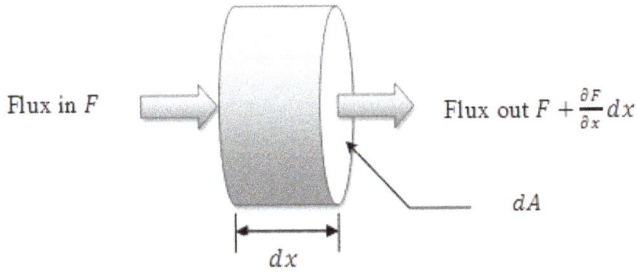

Let F be the mass flux of solute entering in to the control volume per unit area per unit time and dA is the cross sectional area of the control volume.

The mass flux in to the control volume is

FdA

and the flux out from the control volume is

$$\left(F + \frac{\partial F}{\partial x} dx \right) dA$$

Therefore, the net flux = flux in – flux out

$$= FdA - \left(F + \frac{\partial F}{\partial x} dx \right) dA$$

$$= -\frac{\partial F}{\partial x} dx dA$$

As per the law of conservation of mass

Net mass flux is equal to the rate of change in mass stored in control volume in time dt

Net flux is equal to the rate of change in flux in time dt

Therefore,

$$\frac{\partial (CdxdA)}{\partial t} = -\frac{\partial F}{\partial x} dx dA$$

$$\frac{\partial C}{\partial t} = -\frac{\partial F}{\partial x}$$

Putting the above equations together, we have

$$\frac{\partial C}{\partial t} = -\frac{\partial}{\partial x}\left(-D_d\frac{\partial C}{\partial x}\right)$$

Considering diffusivity is independent of solute concentration, we have

$$\frac{\partial C}{\partial t} = -D_d\frac{\partial^2 C}{\partial x^2}$$

This is the Fick's second law of diffusion. This is the governing equation for the system where concentrations are changing with time.

In case of steady state, the equation becomes

$$-\frac{\partial}{\partial x}\left(-D_d\frac{\partial C}{\partial x}\right) = 0$$

Solving,

$$-D_d\frac{\partial C}{\partial x} = F = \text{Constant}$$

This is again the first law of Fick's. Therefore first Fick's law is the simplified format of the second Fick's law when applied to a steady state condition.

In case three-dimensional space, the Fick's second law can be written as,

$$\frac{\partial C}{\partial t} = D_x\frac{\partial^2 C}{\partial x^2} + D_y\frac{\partial^2 C}{\partial x^2} + D_z\frac{\partial^2 C}{\partial z^2}$$

Where, D_x, D_y, and D_z are the diffusion coefficient in x, y, and z directions.

The equation above can be solved using numerical techniques. This equation is also amenable to analytical treatment for particular initial and boundary conditions.

Effective Diffusion Coefficient

In case of flow through porous media, the dispersion process of solute is not as fast as in water because the solute ions have to follow longer pathways when they move through the tortuous path of the porous matrix. As such, an effective diffusion coefficient has to be used to account this phenomenon. The effective diffusion coefficient is defined as,

$$D^* = \omega D_d$$

Where, ω is a coefficient that is related to the tortuosity of the porous matrix. The value of ω is always less than 1 and can be obtained using laboratory study. Freeze and Cherry (1979), based on their experimental studies suggested that ω is in between 0.5 and 0.01. Perkins and Johnson (1963) did a sand column study and found that ω is around 0.7 for sand.

Advection Process

Advection is a process by which the dissolved solids are transported with the flowing groundwater. The amount of solute transported by the advection process is a function of quantity of groundwater flowing and the concentration of solute in the groundwater. If v is the average linear velocity of water in porous matrix, C is the concentration of solute in groundwater and η is the effective porosity of the porous matrix, the one dimensional mass flux due to advection will be,

$$F_x = v\eta C$$

$$Where, v = -\frac{K}{\eta}\frac{dh}{dx}$$

and η is the effective porosity which is the porosity through which flow can actually occur.

Consider the control volume shown in the figure below

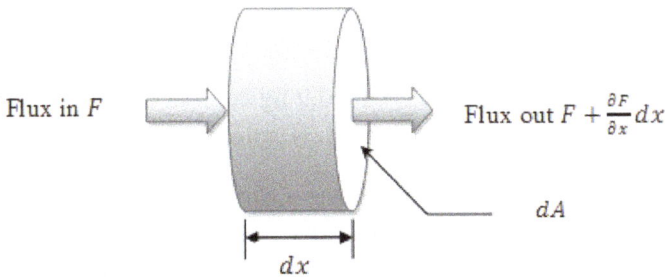

One dimensional control volume

Flux in to the control volume is

FdA

Flux out from the control volume is

$$\left(F + \frac{\partial F}{\partial x}dx \right)dA$$

Net Flux = Flux in – Flux out

$$= FdA - \left(F + \frac{\partial F}{\partial x}dx \right)dA$$

$$= -\frac{\partial F}{\partial x} dx dA$$

As per the law of conservation of mass, net flux is equal to the rate of change in flux in time dt

$$\frac{\partial(C\eta)}{\partial t} dx dA = -\frac{\partial F}{\partial x} dx dA$$

Here, $C\eta dx dA$ is the total volume of solute in the control volume

Putting, $F = v\eta C$

$$\frac{\partial(C\eta)}{\partial t} = -\frac{\partial(v\eta C)}{\partial x}$$

$$\Rightarrow \frac{\partial C}{\partial t} = -v\frac{\partial C}{\partial x} \text{ [Considering } \eta \text{ is a constant, i.e. the aquifer is homogeneous]}$$

This is the one dimensional advection transport equation.

Mechanical Dispersion

In the case of flow through porous media, the solute containing water is not moving at the same velocity as that of water. As a result additional mixing is occurred along the flow path. This additional mixing is called mechanical dispersion. It may be noted that the mixing that occurs along the direction of flow path is called longitudinal dispersion. On the other hand, the mixing that occurs normal to the flow path is called transverse dispersion.

If it is assumed that the phenomenon of mechanical dispersion can be explained by Fick's law, then a coefficient of mechanical dispersion can be introduced. In this case, the coefficient of mechanical dispersion is a function of the average linear velocity. The amount of dispersion in longitudinal direction and transverse direction is different. Therefore, the coefficient in longitudinal direction is called coefficient of longitudinal mechanical dispersion ($\alpha_L v_L$) and that in the transverse direction is called coefficient of transverse mechanical dispersion ($\alpha_T v_L$). Where v_L is the average linear velocity in the longitudinal direction (L/T), α_L is the dynamic dispersivity in the longitudinal direction (L) and α_T is the dynamic dispersivity in the transverse direction (L).

Hydrodynamic Dispersion

In case of flow through porous media, the molecular diffusion process and mechanical dispersion process are difficult to separate. As such, these two processes are combined by a single parameter called hydrodynamic dispersion coefficient which can be expressed as,

$$D_L = D^* + \alpha_L v_L$$

$$D_T = D^* + \alpha_T v_L$$

Where D_L is the longitudinal hydrodynamic dispersion and D_T is the transverse hydrodynamic dispersion

Groundwater Pollution

Groundwater pollution example in Lusaka, Zambia where the pit latrine in the background is polluting the shallow well in the foreground with pathogens and nitrate

Groundwater pollution (also called groundwater contamination) occurs when pollutants are released to the ground and make their way down into groundwater. It can also occur naturally due to the presence of a minor and unwanted constituent, contaminant or impurity in the groundwater, in which case it is more likely referred to as contamination rather than pollution.

The pollutant creates a contaminant plume within an aquifer. Movement of water and dispersion within the aquifer spreads the pollutant over a wider area. Its advancing boundary, often called a plume edge, can intersect with groundwater wells or daylight into surface water such as seeps and spring, making the water supplies unsafe for humans and wildlife. The movement of the plume, called a plume front, may be analyzed through a hydrological transport model or groundwater model. Analysis of groundwater pollution may focus on soil characteristics and site geology, hydrogeology, hydrology, and the nature of the contaminants.

Pollution can occur from on-site sanitation systems, landfills, effluent from wastewater treatment plants, leaking sewers, petrol filling stations or from over application of fertilizers in agriculture. Pollution (or contamination) can also occur from naturally occurring contaminants, such as arsenic or fluoride. Using polluted groundwater causes hazards to public health through poisoning or the spread of disease.

Different mechanisms have influence on the transport of pollutants, e.g. diffusion, adsorption, precipitation, decay, in the groundwater. The interaction of groundwater contamination with surface waters is analyzed by use of hydrology transport models.

Pollutant Types

Contaminants found in groundwater cover a broad range of physical, inorganic chemical, organic chemical, bacteriological, and radioactive parameters. Principally, many of the same pollutants that play a role in surface water pollution may also be found in polluted groundwater, although their respective importance may differ.

Arsenic and Fluoride

Arsenic and fluoride have been recognized by the World Health Organization (WHO) as the most serious inorganic contaminants in drinking-water on a worldwide basis.

The metalloid arsenic can occur naturally in groundwater, as seen most frequently in Asia, including in China, India and Bangladesh. In the Ganges Plain of northern India and Bangladesh severe contamination of groundwater by naturally occurring arsenic affects 25% of water wells in the shallower of two regional aquifers.

Arsenic in groundwater can also be present where there are mining operations or mine waste dumps that will leach arsenic.

Natural fluoride in groundwater is of growing concern as deeper groundwater is being used, "with more than 200 million people at risk of drinking water with elevated concentrations." Fluoride can especially be released from acidic volcanic rocks and dispersed volcanic ash when water hardness is low. High levels of fluoride in groundwater is a serious problem in the Argentinean Pampas, Chile, Mexico, India, Pakistan, the East African Rift, and some volcanic islands (Tenerife)

In areas that have naturally occurring high levels of fluoride in groundwater which is used for drinking water, both dental and skeletal fluorosis can be prevalent and severe.

Pathogens

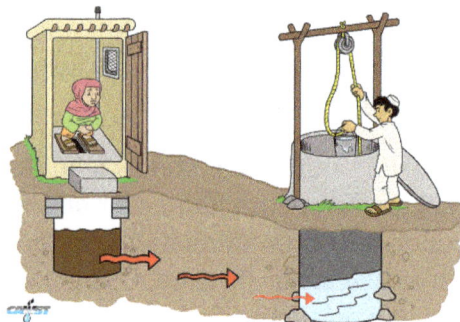

Waterborne diseases can be spread via a groundwater well which is contaminated with fecal pathogens from pit latrines

Pathogens contained in feces can lead to groundwater pollution when they are given the opportunity to reach the groundwater, making it unsafe for drinking. Of the four pathogen types that are present in feces (bacteria, viruses, protozoa and helminths or helminth eggs), the first three can be commonly found in polluted groundwater, whereas the relatively large helminth eggs are usually filtered out by the soil matrix.

Groundwater that is contaminated with pathogens can lead to fatal fecal-oral transmission of diseases (e.g. cholera, diarrhoea).

Nitrate

Nitrate is the most common chemical contaminant in the world's groundwater and aquifers. In some low-income countries nitrate levels in groundwater are extremely high, causing significant health problems. It is also stable (it does not degrade) under high oxygen conditions.

Nitrate levels above 10 mg/L (10 ppm) in groundwater can cause "blue baby syndrome" (acquired methemoglobinemia). Drinking water quality standards in the European Union stipulate less than 50 mg/L for nitrate in drinking water.

However, the linkages between nitrates in drinking water and blue baby syndrome have been disputed in other studies. The syndrome outbreaks might be due to other factors than elevated nitrate concentrations in drinking water.

Elevated nitrate levels in groundwater can be caused by on-site sanitation, sewage sludge disposal and agricultural activities. It can therefore have an urban or agricultural origin.

Organic Compounds

Volatile organic compounds (VOCs) are a dangerous contaminant of groundwater. They are generally introduced to the environment through careless industrial practices. Many of these compounds were not known to be harmful until the late 1960s and it was some time before regular testing of groundwater identified these substances in drinking water sources.

Primary VOC pollutants found in groundwater include aromatic hydrocarbons such as BTEX compounds (benzene, toluene, ethylbenzene and xylenes), and chlorinated solvents including tetrachloroethylene (PCE), trichloroethylene (TCE), and vinyl chloride (VC). BTEX are important components of gasoline. PCE and TCE are industrial solvents historically used in dry cleaning processes and as a metal degreaser, respectively.

Other organic pollutants present in groundwater and derived from industrial operations are the polycyclic aromatic hydrocarbons (PAHs). Due to its molecular weight, Naphthalene is the most soluble and mobile PAH found in groundwater, whereas benzo(a)pyrene is the most toxic one. PAHs are generally produced as byproducts by incomplete combustion of organic matter.

Organic pollutants can also be found in groundwater as insecticides and herbicides. As many other synthetic organic compounds, most pesticides have very complex molecular structures. This complexity determines the water solubility, adsorption capacity, and mobility of pesticides in the groundwater system. Thus, some types of pesticides are more mobile than others so they can more easily reach a drinking-water source.

Metals

Several trace metals can occurs naturally in some rocks or enter in the environmental from natural processes such as weathering. However, industrial activities such as mining and metallurgy, solid waste disposal, painting and enamel works can lead to severe groundwater pollution with elevated concentrations of toxic metals including lead, cadmium and chromium.

The migration of metals (and metalloids) in groundwater will be affected by several factors, in particular by chemical reactions which determine the partitioning of contaminants among different phases and species. Thus, the mobility of metals primarily depends on the pH and redox state of groundwater.

Others

Other organic pollutants include a range of organohalides and other chemical compounds, petroleum hydrocarbons, various chemical compounds found in personal hygiene and cosmetic products, drug pollution involving pharmaceutical drugs and their metabolites. Inorganic pollutants might include other nutrients such as ammonia and phosphate, and radionuclides such as uranium (U) or radon (Rn) naturally present in some geological formations. Saltwater intrusion is also an example of natural contamination, but is very often intensified by human activities.

Groundwater pollution is a worldwide issue. A study of the groundwater quality of the principal aquifers of the United States conducted between 1991 and 2004, showed that 23% of domestic wells had contaminants at levels greater than human-health benchmarks. Another study suggested that the major groundwater pollution problems in Africa, considering the order of importance are: (1) nitrate pollution, (2) pathogenic agents, (3) organic pollution, (4) salinization, and (5) acid mine drainage.

Causes

Naturally-occurring (Geogenic)

"Geogenic" refers to naturally occurring as a result from geological processes.

The natural arsenic pollution occurs because aquifer sediments contain organic matter that generates anaerobic conditions in the aquifer. These conditions result in the microbial dissolution of iron oxides in the sediment and, thus, the release of the arsenic, normally strongly bound to iron oxides, into the water. As a consequence, arsenic-rich groundwater is often iron-rich, although secondary processes often obscure the association of dissolved arsenic and dissolved iron. Arsenic is found in groundwater most commonly as the reduced species arsenite and the oxidized species arsenate, being the acute toxicity of arsenite somewhat greater than that of arsenate. Investigations by WHO indicated that 20% of 25,000 boreholes tested in Bangladesh had arsenic concentrations exceeding 50 µg/l.

The occurrence of fluoride is close related to the abundance and solubility of fluoride-containing minerals such as fluorite (CaF_2). Considerably high concentrations of fluoride in groundwater are typically caused by a lack of calcium in the aquifer. Health problems associated with dental fluorosis may occur when fluoride concentrations in groundwater exceed 1.5 mg/l, which is the WHO guideline value since 1984.

The Swiss Federal Institute of Aquatic Science and Technology (EAWAG) has recently developed the interactive Groundwater Assessment Platform (GAP), where the geogenic risk of contamination in a given area can be estimated using geological, topographical and other environmental data without having to test samples from every single groundwater resource. This tool also allows the user to produce probability risk mapping for both arsenic and fluoride.

High concentrations of parameters like salinity, iron, manganese, uranium, radon and chromium, in groundwater, may also be of geogenic origin. This contaminants can be important locally but they are not as widespread as arsenic and fluoride.

On-site Sanitation Systems

A traditional housing compound near Herat, Afghanistan, where a shallow water supply well (foreground) is in close proximity to the pit latrine (behind the white greenhouse) leading to contamination of the groundwater

Groundwater pollution with pathogens and nitrate can also occur from the liquids infiltrating into the ground from on-site sanitation systems such as pit latrines and septic tanks, depending on the population density and the hydrogeological conditions.

Factors controlling the fate and transport of pathogens are quite complex and the interaction among them is not well understood. If the local hydrogeological conditions (which can vary within a space of a few square kilometres) are ignored, simple on-site sanitation infrastructures such as pit latrines can cause significant public health risks via contaminated groundwater.

Liquids leach from the pit and pass the unsaturated soil zone (which is not completely filled with water). Subsequently, these liquids from the pit enter the groundwater where they may lead to groundwater pollution. This is a problem if a nearby water well is used to supply groundwater for drinking water purposes. During the passage in the soil, pathogens can die off or be adsorbed significantly, mostly depending on the travel time between the pit and the well. Most, but not all pathogens die within 50 days of travel through the subsurface.

The degree of pathogen removal strongly varies with soil type, aquifer type, distance and other environmental factors. For example, the unsaturated zone becomes "washed" during extended periods of heavy rain, providing hydraulic pathway for the quick pass of pathogens. It is difficult to estimate the safe distance between a pit latrine or a septic tank and a water source. In any case, such recommendations about the safe distance are mostly ignored by those building pit latrines. In addition, household plots are of a limited size and therefore pit latrines are often built much closer to groundwater wells than what can be regarded as safe. This results in groundwater pollution and household members falling sick when using this groundwater as a source of drinking water.

Sewage (Treated and Untreated)

Groundwater pollution can be caused by untreated waste discharge leading to diseases like skin lesions, bloody diarrhea and dermatitis. This is more common in locations having limited wastewater treatment infrastructure, or where there are systematic failures of the on-site sewage disposal system. Along with pathogens and nutrients, untreated sewage can also have an important load of heavy metals that may seep into the groundwater system.

The treated effluent from sewage treatment plants may also reach the aquifer if the effluent is infiltrated or discharged to local surface water bodies. Therefore, those substances that are not removed in conventional sewage treatment plants may reach the groundwater as well. For example, detected concentrations of pharmaceutical residues in groundwater were in the order of 50 ng/L in several locations in Germany. This is because in conventional sewage treatment plants, micro-pollutants such as hormones, pharmaceutical residues and other micro-pollutants contained in urine and feces are only partially removed and the remainder is discharged into surface water, from where it may also reach the groundwater.

Groundwater pollution can also occur from leaking sewers which has been observed for example in Germany. This can also lead to potential cross-contamination of drinking-water supplies.

Spreading wastewater or sewage sludge in agriculture may also be included as sources of faecal contamination in groundwater.

Fertilizers and Pesticides

Nitrate can also enter the groundwater via excessive use of fertilizers, including manure spreading. This is because only a fraction of the nitrogen-based fertilizers is converted to produce and other plant matter. The remainder accumulates in the soil or lost as run-off. High application rates of nitrogen-containing fertilizers combined with the high water-solubility of nitrate leads to increased runoff into surface water as well as leaching into groundwater, thereby causing groundwater pollution. The excessive use of nitrogen-containing fertilizers (be they synthetic or natural) is particularly damaging, as much of the nitrogen that is not taken up by plants is transformed into nitrate which is easily leached.

Poor management practices in manure spreading can introduce both
pathogens and nutrients (nitrate) in the groundwater system

The nutrients, especially nitrates, in fertilizers can cause problems for natural habitats and for human health if they are washed off soil into watercourses or leached through soil into groundwater. The heavy use of nitrogenous fertilizers in cropping systems is the largest contributor to anthropogenic nitrogen in groundwater worldwide.

Feedlots/animal corrals can also lead to the potential leaching of nitrogen and metals to groundwater. Over application of animal manure may also result in groundwater pollution with pharmaceutical residues derived from veterinary drugs.

The US Environmental Protection Agency (EPA) and the European Commission are seriously dealing with the nitrate problem related to agricultural development, as a major water supply problem that requires appropriate management and governance.

Runoff of pesticides may leach into groundwater causing human health problems from contaminated water wells. Pesticide concentrations found in groundwater are typically low, and often the regulatory human health-based limits exceeded are also very low. The organophosphorus insecticide monocrotophos (MCP) appears to be one of a few hazardous, persistent, soluble and mobile (it does not bind with minerals in soils) pesticides able to reach a drinking-water source. In general, more pesticide compounds are being detected as groundwater quality monitoring programs have become more extensive; however, much less monitoring has been conducted in developing countries due to the high analysis costs.

Commercial and Industrial Leaks

A wide variety of both inorganic and organic pollutants have been found in aquifers underlying commercial and industrial activities.

Ore mining and metal processing facilities are the primary responsible of the presence of metals in groundwater of anthropogenic origin, including arsenic. The low pH associated with acid mine drainage (AMD) contributes to the solubility of potential toxic metals that can eventually enter the groundwater system.

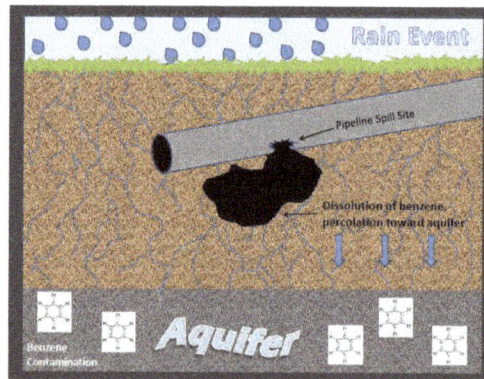

Oil spills associated with underground pipelines and tanks can release benzene and other soluble petroleum hydrocarbons that rapidly percolate down into the aquifer

There is a increasing concern over the groundwater pollution by gasoline leaked from petroleum underground storage tanks (USTs) of gas stations. BTEX compounds are the most common additives of the gasoline. BTEX compounds, including benzene, have densities lower than water (1 g/ml). Similar to the oil spills on the sea, the non-miscible phase, referred to as Light Non-Aqueous Phase Liquid (LNAPL), will "float" upon the water table in the aquifer.

Chlorinated solvents are used in nearly any industrial practice where degreasing removers are required. PCE is a highly utilized solvent in the dry cleaning industry because of its cleaning effectiveness and relatively low cost. It has also been used for metal-degreasing operations. Because it is is highly volatile, it is more frequently found in groundwater than in surface water. TCE has historically been used as a metal cleaning.

The military facility Anniston Army Dept (ANAD) in the United States was placed on the US EPA Superfund National Priorities List (NPL) because of groundwater contamination with as much as 27 million pounds of TCE. Both PCE and TCE may degrade to vinyl chloride (VC), the most toxic chlorinated hydrocarbon.

Many types of solvents may have also been disposed illegally, leaking over time to the groundwater system.

Chlorinated solvents such as PCE and TCE have densities higher than water and the non-miscible phase is referred to as Dense Non-Aqueous Phase Liquids (DNAPL). Once they reach the aquifer, they will "sink" and eventually accumulate on the top of low-permeability layers.

Historically, wood-treating facilities have also release insecticides such as pentachlorophenol (PCP) and creosote into the environment, impacting the groundwater resources. PCP is a a highly soluble and toxic obsolete pesticide recently listed in the Stockholm Convention on Persistent Organic Pollutants. PAHs and other semi-VOCs are the common contaminants associated with creosote.

Although non-miscible, both LNAPLs and DNAPLs still have the potential to slowly dissolve into the aqueous (miscible) phase to create a plume and thus become a long-term source of contamination. DNAPLs (chlorinated solvents, heavy PAHs, creosote, PCBs) tend to be difficult to manage as they can reside very deep in the groundwater system.

Hydraulic Fracturing

The recent growth of Hydraulic Fracturing ("Fracking") wells in the United States has raised concerns regarding its potential risks of contaminating groundwater resources. The Environmental Protection Agency (EPA), along with many other researchers, has been delegated to study the relationship between hydraulic fracturing and drinking water resources. While it is possible to perform hydraulic fracturing without having a relevant impact on groundwater resources if stringent controls und quality management measures are in place, there are a number of cases where groundwater pollution due to improper handling or technical failures was observed.

While the EPA has not found significant evidence of a widespread, systematic impact on drinking water by hydraulic fracturing, this may be due to insufficient systematic pre- and post- hydraulic fracturing data on drinking water quality, and the presence of other agents of contamination that preclude the link between shale oil/gas extraction and its impact.

Despite the EPA's lack of profound widespread evidence, other researchers have made significant observations of rising groundwater contamination in close proximity to major shale oil/gas drilling sites located in MarcellusEllsworth, William (2013). "Injection-Induced Earthquakes". Science AAAS.</ref> (British Columbia, Canada). Within one kilometer of these specific sites, a subset of shallow drinking water consistently showed higher concentration levels of methane, ethane, and propane concentrations than normal. An evaluation of higher Helium and other noble gas concentration along with the rise of hydrocarbon levels supports the distinction between hydraulic fracturing fugitive gas and naturally occurring "background" hydrocarbon content. This contamination is speculated to be the result of leaky, failing, or improperly installed gas well casings.

Furthermore, it is theorized that contamination could also result from the capillary migration of deep residual hyper-saline water and hydraulic fracturing fluid, slowly flowing through faults and fractures until finally making contact with groundwater resources; however, many researchers argue that the permeability of rocks overlying shale formations are too low to allow this to ever happen sufficiently. To ultimately prove this theory, there would have to be traces of toxic trihalomethanes (THM) since they are often associated with the presence of stray gas contamination, and typically co-occur with high halogen concentrations in hyper-saline waters. Besides, highly saline waters are a common natural feature in deep groundwater systems.

While conclusions regarding groundwater pollution as the result to hydraulic fracturing fluid flow is restricted in both space and time, researchers have hypothesized that the potential for systematic stray gas contamination depends mainly on the integrity of the shale oil/gas well structure, along with its relative geological location to local fracture systems that could potentially provide flow paths for fugitive gas migration.

Though widespread, systematic contamination by hydraulic fracturing has been heavily disputed, one major source of contamination that has the most consensus among researchers of being the most problematic is site-specific accidental spillage of hydraulic fracturing fluid and produced water. So far, a significant majority of groundwater contamination events are derived from surface-level anthropogenic routes rather than the subsurface flow from underlying shale formations. Examples of such events include: a fracking fluid spillage in Acorn Fork Creek, Kentucky that caused a widespread death among aquatic species in 2007; a 420,000 gallon spillage of hyper-saline produced water that turned a once very-fertile farmland in New Mexico into a dead-zone in 2010; and a 42,000 gallon fracking fluid spillage in Arlington, Texas that necessitated an evacuation of over a 100 homes in 2015. While the damage can be obvious, and much more effort is being done to prevent these accidents from occurring so frequently, the lack of data from fracking oil spills continue to leave researchers in the dark. In many of these events, the data acquired from the leakage or spillage is often very vague, and thus would lead researchers to lacking conclusions.

Researchers from the Federal Institute for Geosciences and Natural Resources (BGR) conducted a modelling study for a deep shale-gas formation in the North German Basin. They concluded that the probability is small that the rise of fracking fluids through the geological underground to the surface will impact shallow groundwater.

Landfill Leachate

Leachate from sanitary landfills can lead to groundwater pollution.

Love Canal was one of the most widely known examples of groundwater pollution. In 1978, residents of the Love Canal neighborhood in upstate New York noticed high rates of cancer and an alarming number of birth defects. This was eventually traced to organic solvents and dioxins from an industrial landfill that the neighborhood had been built over and around, which had then infiltrated into the water supply and evaporated in basements to further contaminate the air. Eight hundred families were reimbursed for their homes and moved, after extensive legal battles and media coverage.

Other

Further causes of groundwater pollution are chemical spills from commercial or industrial operations, chemical spills occurring during transport (e.g. spillage of diesel fuels), illegal waste dumping, infiltration from urban runoff or mining operations, road salts, de-icing chemicals from airports and even atmospheric contaminants since groundwater is part of the hydrologic cycle.

The burial of corpses and their subsequent degradation may also pose a risk of pollution to groundwater.

Mechanisms

The passage of water through the subsurface can provide a reliable natural barrier to contamination but it only works under favorable conditions.

The stratigraphy of the area plays an important role in the transport of pollutants. An area can have layers of sandy soil, fractured bedrock, clay, or hardpan. Areas of karst topography on limestone bedrock are sometimes vulnerable to surface pollution from groundwater. Earthquake faults can also be entry routes for downward contaminant entry. Water table conditions are of great importance for drinking water supplies, agricultural irrigation, waste disposal (including nuclear waste), wildlife habitat, and other ecological issues.

Interactions with Surface Water

Although interrelated, surface water and groundwater have often been studied and managed as separate resources. Surface water seeps through the soil and becomes groundwater. Conversely, groundwater can also feed surface water sources. Sources of surface water pollution are generally grouped into two categories based on their origin.

Interactions between groundwater and surface water are complex. Consequently, groundwater pollution, sometimes referred to as groundwater contamination, is not as easily classified as surface water pollution. By its very nature, groundwater aquifers are susceptible to contamination from sources that may not directly affect surface water bodies, and the distinction of point vs. non-point source may be irrelevant. A spill or ongoing release of chemical or radionuclide contaminants into soil (located away from a surface water body) may not create point or non-point source pollution but can contaminate the aquifer below, creating a toxic plume.

Prevention

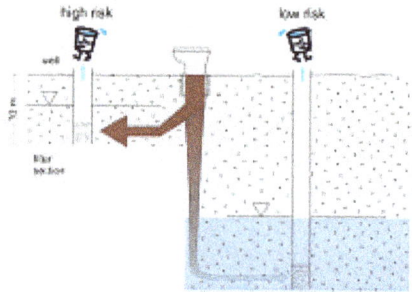

Schematic showing that there is a lower risk of groundwater
pollution with greater depth of the water well

Precautionary Principle

The precautionary principle, evolved from Principle 15 of the Rio Declaration on Environment and Development, is important in protecting groundwater resources from pollution. The precautionary principle provides that "where there are threats of irreversible damage, lack of full scientific certainty shall not be used as reason for postponing cost-effective measures to prevent environmental degradation".

One of the six basic principles of the European Union (EU) water policy is the application of the precautionary principle.

Groundwater Quality Monitoring

Groundwater quality monitoring programs have been implemented regularly in many countries around the world. They are important components to understand the hydrogeological system, and for the development of conceptual models and aquifer vulnerability maps.

Groundwater quality must be regularly monitored across the aquifer to determine trends. Effective groundwater monitoring should be driven by a specific objective, for example, a specific contaminant of concern. Contaminant levels can be compared to the World Health Organization (WHO) guidelines for drinking-water quality. It is not rare that limits of contaminants are reduced as more medical experience is gained.

Sufficient investment should be given to continue monitoring over the long term. When a problem is found, action should be taken to correct it. Waterborne outbreaks in the United States decreased with the introduction of more stringent monitoring (and treatment) requirements in the early 90s.

The community can also help monitor the groundwater quality.

Land Zoning for Groundwater Protection

The development of land-use zoning maps has been implemented by several water authorities at different scales around the world. There are two types of zoning maps: aquifer vulnerability maps and source protection maps.

Aquifer Vulnerability Map

It refers to the intrinsic (or natural) vulnerability of a groundwater system to pollution. Intrinsically, some aquifers are more vulnerable to pollution than other aquifers. Shallow unconfined aquifers are more at risk of pollution because there are fewer layers to filter out contaminants.

The unsaturated zone can play an important role in retarding (and in some cases eliminating) pathogens and so must be considered when assessing aquifer vulnerability. The biological activity is greatest in the top soil layers where the attenuation of pathogens is generally most effective.

Preparation of the vulnerability maps typically involves overlaying several thematic maps of physical factors that have been selected to describe the aquifer vulnerability. The index-based parametric mapping method GOD developed by Foster and Hirata (1988) uses three generally available or readily estimated parameters, the degree of Groundwater hydraulic confinement, geological nature of the Overlying strata and Depth to groundwater. A further approach developed by the US EPA named DRASTIC employs seven hydrogeological factors to develop an index of vulnerability: Depth to water table, net Recharge, Aquifer media, Soil media, Topography (slope), Impact on the vadose zone, and hydraulic Conductivity.

There is a particular debate among hydrogeologist whether aquifer vulnerability should be established in a general (intrinsic) way for all contaminants, or specifically for each pollutant.

Source Protection Map

It refers to the capture areas around an individual groundwater source, such as a water well or a spring, to especially protect them from pollution. Thus, potential sources of degradable pollutants, such as pathogens, can be located at distances which travel times along the flowpaths are long enough for the pollutant to be eliminated through filtration or adsorption.

Analytical methods using equations to define groundwater flow and contaminant transport are the most widely used. The WHPA is a semi-analytical groundwater flow simulation program developed by the US EPA for delineating capture zones in a wellhead protection area.

The simplest form of zoning employs fixed-distance methods where activities are excluded within a uniformly applied specified distance around abstraction points.

Locating on-site Sanitation Systems

As the health effects of most toxic chemicals arise after prolonged exposure, risk to health from chemicals is generally lower than that from pathogens. Thus, the quality of the source protection measures is an important component in controlling whether pathogens may be present in the final drinking-water.

On-site sanitation systems can be designed in such a way that groundwater pollution from these sanitation systems is prevented from occurring. Detailed guidelines have been developed to estimate safe distances to protect groundwater sources from pollution from on-site sanitation. The following criteria have been proposed for safe siting (i.e. deciding on the location) of on-site sanitation systems:

- Horizontal distance between the drinking water source and the sanitation system

    o Guideline values for horizontal separation distances between on-site sanitation systems and water sources vary widely (e.g. 15 to 100 m horizontal distance between pit latrine and groundwater wells)

- Vertical distance between drinking water well and sanitation system

- Aquifer type

- Groundwater flow direction

- Impermeable layers

- Slope and surface drainage

- Volume of leaking wastewater

- Superposition, i.e. the need to consider a larger planning area

As a very general guideline it is recommended that the bottom of the pit should be at least 2 m above groundwater level, and a minimum horizontal distance of 30 m between a pit and a water source is normally recommended to limit exposure to microbial contamination. However, no general statement should be made regarding the minimum lateral separation distances required to prevent contamination of a well from a pit latrine. For example, even 50 m lateral separation distance might not be sufficient in a strongly karstified system with a downgradient supply well or spring, while 10 m lateral separation distance is completely sufficient if there is a well developed clay cover layer and the annular space of the groundwater well is well sealed.

Legislation

Institutional and legal issues are critical in determining the success or failure of ground-water protection policies and strategies.

Sign near Mannheim, Germany indicating a zone as a dedicated "groundwater protection zone"

United States

In November 2006, the Environmental Protection Agency published the Ground Water Rule in the United States Federal Register. The EPA was worried that the ground water system would be vulnerable to contamination from fecal matter. The point of the rule was to keep microbial pathogens out of public water sources. The 2006 Ground Water Rule was an amendment of the 1996 Safe Drinking Water Act.

The ways to deal with groundwater pollution that has already occurred can be grouped into the following categories: containing the pollutants to prevent them from migrating further; removing the pollutants from the aquifer; remediating the aquifer by either immobilizing or detoxifying the contaminants while they are still in the aquifer (in-situ); treating the groundwater at its point of use; or abandoning the use of this aquifer's groundwater and finding an alternative source of water.

Management

Point-of-use Treatment

Portable water purification devices or "point-of-use" (POU) water treatment systems and field water disinfection techniques can be used to remove some forms of groundwater pollution prior to drinking, namely any fecal pollution. Many commercial portable water purification systems or chemical additives are available which can remove pathogens, chlorine, bad taste, odors, and heavy metals like lead and mercury.

Techniques include boiling, filtration, activated charcoal absorption, chemical disinfection, ultraviolet purification, ozone water disinfection, solar water disinfection, solar distillation, homemade water filters.

Arsenic removal filters (ARF) are dedicated technologies typically installed to remove arsenic. Many of these technologies require a capital investment and long-term maintenance. Filters in Bangladesh are usually abandoned by the users due to their high cost and complicated maintenance, which is also quite expensive.

Groundwater Remediation

Groundwater pollution is much more difficult to abate than surface pollution because groundwater can move great distances through unseen aquifers. Non-porous aquifers such as clays partially purify water of bacteria by simple filtration (adsorption and absorption), dilution, and, in some cases, chemical reactions and biological activity; however, in some cases, the pollutants merely transform to soil contaminants. Groundwater that moves through open fractures and caverns is not filtered and can be transported as easily as surface water. In fact, this can be aggravated by the human tendency to use natural sinkholes as dumps in areas of karst topography.

Pollutants and contaminants can be removed from ground water by applying various techniques thereby making it safe for use. Ground water treatment (or remediation) techniques span biological, chemical, and physical treatment technologies. Most ground water treatment techniques utilize a combination of technologies. Some of the biological treatment techniques include bioaugmentation, bioventing, biosparging, bioslurping, and phytoremediation. Some chemical treatment techniques include ozone and oxygen gas injection, chemical precipitation, membrane separation, ion exchange, carbon absorption, aqueous chemical oxidation, and surfactant enhanced recovery. Some chemical techniques may be implemented using nanomaterials. Physical treatment techniques include, but are not limited to, pump and treat, air sparging, and dual phase extraction.

Abandonment

If treatment or remediation of the polluted groundwater is deemed to be too difficult or expensive then abandoning the use of this aquifer's groundwater and finding an alternative source of water is the only other option.

Society and culture

Hinkley, U.S.

The town of Hinkley, California (U.S.), had its groundwater contaminated with hexavalent chromium starting in 1952, resulting in a legal case against Pacific Gas & Electric (PG&E) and a multimillion-dollar settlement in 1996. The legal case was dramatized in the film Erin Brockovich, released in 2000.

California, U.S.

Nitrates and water contamination in California's Central Valley

Walkerton, Canada

In the year 2000, groundwater pollution occurred in the small town of Walkerton, Canada leading to seven deaths in what is known as the Walkerton E. Coli outbreak. The water supply which was drawn from groundwater became contaminated with the highly dangerous O157:H7 strain of E. coli bacteria. This contamination was due to farm runoff into an adjacent water well that was vulnerable to groundwater pollution.

Lusaka, Zambia

The peri-urban areas of Lusaka, the capital of Zambia, have ground conditions which are strongly karstified and for this reason – together with the increasing population density in these peri-urban areas – pollution of water wells from pit latrines is a major public health threat there.

Advection-dispersion Equation

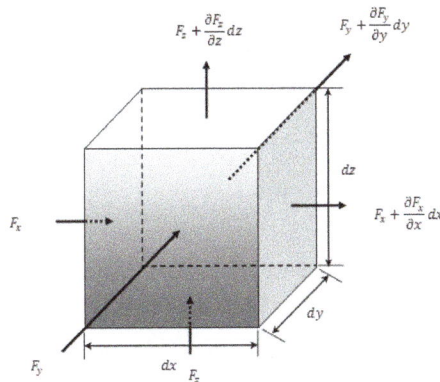

Elementary control volume

In this chapter, we will derive the advection-dispersion equation for solute transport in porous medium.

Consider the small representative elementary volume as shown in the figure above where a mass of solute is transported by advection and hydrodynamic dispersion processes.

Solute entering into the control volume in x direction due to advection is

$v_x \eta C dydz$

Solute entering to the control volume in x direction due to dispersion is

$$-\eta D_x \frac{\partial C}{\partial x} dydz$$

Where, v_x average linear velocity in x direction, n is the porosity of the porous matrix, C is the solute concentration, D_x is the hydrodynamic dispersion coefficient for x direction. The negative sign indicates that the solute mass flows from higher concentration to lower concentration .

Now, total mass of solute transported per unit cross sectional area by advection and dispersion in xdirection is

$$F_x = \left(v_x \eta C dydz - \eta D_x \frac{\partial C}{\partial x} dydz \right) / dydz$$

$$F_x = \left(v_x \eta C - \eta D_x \frac{\partial C}{\partial x} \right)$$

Similarly, total mass of solute transported per unit cross sectional area by advection and dispersion in y direction is

$$F_y = \left(v_y \eta C - \eta D_y \frac{\partial C}{\partial y} \right)$$

Similarly, total mass of solute transported per unit cross sectional area by advection and dispersion in z direction is

$$F_z = \left(v_z \eta C - \eta D_z \frac{\partial C}{\partial z} \right)$$

Total mass of solute entering the representative elementary volume in x direction is

$F_x dydz$

Total mass of solute leaving the representative elementary volume in x direction is

$$\left(F_x + \frac{\partial F_x}{\partial x} dx \right) dydz$$

Net solute mass flux in x direction is

$$F_x dydz - \left(F_x + \frac{\partial F_x}{\partial x} dx \right) dydz = -\frac{\partial F_x}{\partial x} dxdydz$$

Similarly, net solute mass flux in y direction is

$$-\frac{\partial F_y}{\partial y} dxdydz$$

And, net solute mass flux in z direction is

$$-\frac{\partial F_y}{\partial z} dxdydz$$

Total net flux of the representative elementary volume is

$$\frac{\partial F_x}{\partial x} dxdydz - \frac{\partial F_y}{\partial y} dxdydz - \frac{\partial F_z}{\partial z} dxdydz$$

$$= -\left(\frac{\partial F_x}{\partial x} + \frac{\partial F_y}{\partial y} + \frac{\partial F_z}{\partial z} \right) dxdydz$$

As per the law of conservation of mass, net flux of the representative elementary volume is equal to the rate of mass change in the representative elementary volume. The rate of mass change in the representative elementary volume is

$$\eta \frac{\partial C}{\partial t} dxdydz$$

Equating,

$$-\left(\frac{\partial F_x}{\partial x} + \frac{\partial F_y}{\partial y} + \frac{\partial F_z}{\partial z} \right) dxdydz = \eta \frac{\partial C}{\partial t} dxdydz$$

$$\Rightarrow -\left(\frac{\partial F_x}{\partial x} + \frac{\partial F_y}{\partial y} + \frac{\partial F_z}{\partial z} \right) \eta \frac{\partial C}{\partial t}$$

$$\Rightarrow -\left[\frac{\partial}{\partial x} \left(v_x \eta C - \eta D_x \frac{\partial C}{\partial x} \right) + \frac{\partial}{\partial y} \left(v_y \eta C - \eta D_y \frac{\partial C}{\partial y} \right) \right] + \frac{\partial}{\partial z} \left(v_z \eta C - \eta D_z \frac{\partial C}{\partial t} \right) = \eta \frac{\partial C}{\partial t}$$

Considering η is a constant, we have

$$\Rightarrow \frac{\partial}{\partial x}\left(D_x \frac{\partial C}{\partial x}\right) + \frac{\partial}{\partial y}\left(D_y \frac{\partial C}{\partial y}\right) + \frac{\partial}{\partial z}\left(D_z \frac{\partial C}{\partial t}\right) - \frac{\partial}{\partial x}(v_x C) - \frac{\partial}{\partial y}(v_y C) - \frac{\partial}{\partial z}(v_z C) = \frac{\partial C}{\partial t}$$

This the advection-dispersion equation for conservative solute transport in porous media. Conservative solute means that the solute does not interact with the porous media or it does not undergo biological or radioactive decay.

In case of homogeneous medium with uniform velocity field

$$D_x \frac{\partial^2 C}{\partial x^2} + D_y \frac{\partial^2 C}{\partial y^2} + D_z \frac{\partial^2 C}{\partial z^2} - v_x \frac{\partial C}{\partial x} - v_y \frac{\partial C}{\partial y} - v_z \frac{\partial C}{\partial z} = \frac{\partial C}{\partial t}$$

In case of one dimensional flow in homogeneous medium, the equation becomes,

$$D_x \frac{\partial^2 C}{\partial x^2} - v_x \frac{\partial C}{\partial x} = \frac{\partial C}{\partial t}$$

Advection

In physics, engineering, and earth sciences, advection is the transport of a substance by bulk motion. The properties of that substance are carried with it. Generally the majority of the advected substance is a fluid. The properties that are carried with the advected substance are conserved properties such as energy. An example of advection is the transport of pollutants or silt in a river by bulk water flow downstream. Another commonly advected quantity is energy or enthalpy. Here the fluid may be any material that contains thermal energy, such as water or air. In general, any substance or conserved, extensive quantity can be advected by a fluid that can hold or contain the quantity or substance.

During advection, a fluid transports some conserved quantity or material via bulk motion. The fluid's motion is described mathematically as a vector field, and the transported material is described by a scalar field showing its distribution over space. Advection requires currents in the fluid, and so cannot happen in rigid solids. It does not include transport of substances by molecular diffusion.

Advection is sometimes confused with the more encompassing process of convection which is the combination of advective transport and diffusive transport.

In meteorology and physical oceanography, advection often refers to the transport of some property of the atmosphere or ocean, such as heat, humidity or salinity. Advection is important for the formation of orographic clouds and the precipitation of water from clouds, as part of the hydrological cycle.

Distinction Between Advection and Convection

The term advection sometimes serves as a synonym for convection, but technically, convection covers the sum of transport both by diffusion and by advection. Advective transport describes the movement of some quantity via the bulk flow of a fluid (as in a river or pipeline).

Meteorology

In meteorology and physical oceanography, advection often refers to the horizontal transport of some property of the atmosphere or ocean, such as heat, humidity or salinity, and convection generally refers to vertical transport (vertical advection). Advection is important for the formation of orographic clouds (terrain-forced convection) and the precipitation of water from clouds, as part of the hydrological cycle.

Other Quantities

The advection equation also applies if the quantity being advected is represented by a probability density function at each point, although accounting for diffusion is more difficult.

Mathematics of Advection

The advection equation is the partial differential equation that governs the motion of a conserved scalar field as it is advected by a known velocity vector field. It is derived using the scalar field's conservation law, together with Gauss's theorem, and taking the infinitesimal limit.

One easily visualized example of advection is the transport of ink dumped into a river. As the river flows, ink will move downstream in a "pulse" via advection, as the water's movement itself transports the ink. If added to a lake without significant bulk water flow, the ink would simply disperse outwards from its source in a diffusive manner, which is not advection. Note that as it moves downstream, the "pulse" of ink will also spread via diffusion. The sum of these processes is called convection.

The Advection Equation

In Cartesian coordinates the advection operator is

$$\mathbf{u} \cdot \nabla = u_x \frac{\partial}{\partial x} + u_y \frac{\partial}{\partial y} + u_z \frac{\partial}{\partial z}.$$

where $\mathbf{u} = (u_x, u_y, u_z)$ is the velocity field, and ∇ is the del operator (note that Cartesian coordinates are used here).

The advection equation for a conserved quantity described by a scalar field ψ is expressed mathematically by a continuity equation:

$$\frac{\partial \psi}{\partial t} + \nabla \cdot (\psi \mathbf{u}) = 0$$

where $\nabla \cdot$ is the divergence operator and again $\mathbf{u}$ is the velocity vector field. Frequently, it is assumed that the flow is incompressible, that is, the velocity field satisfies

$$\nabla \cdot \mathbf{u} = 0$$

and $\mathbf{u}$ is said to be solenoidal. If this is so, the above equation can be rewritten as

$$\frac{\partial \psi}{\partial t} + \mathbf{u} \cdot \nabla \psi = 0.$$

In particular, if the flow is steady, then

$$\mathbf{u} \cdot \nabla \psi = 0$$

which shows that ψ is constant along a streamline. Hence, $\partial \psi / \partial t = 0$, so ψ doesn't vary in time.

If a vector quantity $\mathbf{a}$ (such as a magnetic field) is being advected by the solenoidal velocity field $\mathbf{u}$, the advection equation above becomes:

$$\frac{\partial \mathbf{a}}{\partial t} + (\mathbf{u} \cdot \nabla) \mathbf{a} = 0.$$

Here, $\mathbf{a}$ is a vector field instead of the scalar field ψ.

Solving the Equation

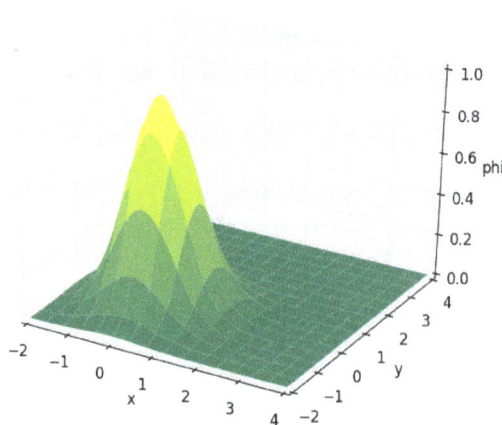

A simulation of the advection equation where u = (sin t, cos t) is solenoidal

The advection equation is not simple to solve numerically: the system is a hyperbolic partial differential equation, and interest typically centers on discontinuous "shock" solutions (which are notoriously difficult for numerical schemes to handle).

Even with one space dimension and a constant velocity field, the system remains difficult to simulate. The equation becomes

$$\frac{\partial \psi}{\partial t} + u_x \frac{\partial \psi}{\partial x} = 0$$

where $\psi = \psi(x,t)$ is the scalar field being advected and u_x is the x component of the vector $\mathbf{u} = (u_x, 0, 0)$.

Treatment of the Advection Operator in the Incompressible Navier Stokes Equations

According to Zang, numerical simulation can be aided by considering the skew symmetric form for the advection operator.

$$\frac{1}{2}\mathbf{u} \cdot \nabla \mathbf{u} + \frac{1}{2}\nabla(\mathbf{uu})$$

where

$$\nabla(\mathbf{uu}) = [\nabla(\mathbf{u}u_x), \nabla(\mathbf{u}u_y), \nabla(\mathbf{u}u_z)]$$

and $\mathbf{u}$ is the same as above.

Since skew symmetry implies only imaginary eigenvalues, this form reduces the "blow up" and "spectral blocking" often experienced in numerical solutions with sharp discontinuities.

Using vector calculus identities, these operators can also be expressed in other ways, available in more software packages for more coordinate systems.

$$\mathbf{u} \cdot \nabla \mathbf{u} = \nabla\left(\frac{\|\mathbf{u}\|^2}{2}\right) + (\nabla \times \mathbf{u}) \times \mathbf{u}$$

$$\frac{1}{2}\mathbf{u} \cdot \nabla \mathbf{u} + \frac{1}{2}\nabla(\mathbf{uu}) = \nabla\left(\frac{\|\mathbf{u}\|^2}{2}\right) + (\nabla \times \mathbf{u}) \times \mathbf{u} + \frac{1}{2}\mathbf{u}(\nabla \cdot \mathbf{u})$$

This form also makes visible that the skew symmetric operator introduces error when the velocity field diverges. Solving the advection equation by numerical methods is very challenging and there is a large scientific literature about this.

Dispersion (Water Waves)

In fluid dynamics, dispersion of water waves generally refers to frequency dispersion, which means that waves of different wavelengths travel at different phase speeds. Water waves, in this context, are waves propagating on the water surface, with gravity and surface tension as the restoring forces. As a result, water with a free surface is generally considered to be a dispersive medium.

For a certain water depth, surface gravity waves – i.e. waves occurring at the air–water interface and gravity as the only force restoring it to flatness – propagate faster with increasing wavelength. On the other hand, for a given (fixed) wavelength, gravity waves in deeper water have a larger phase speed than in shallower water. In contrast with the behavior of gravity waves, capillary waves (i.e. only forced by surface tension) propagate faster for shorter wavelengths.

Besides frequency dispersion, water waves also exhibit amplitude dispersion. This is a nonlinear effect, by which waves of larger amplitude have a different phase speed from small-amplitude waves.

Frequency Dispersion for Surface Gravity Waves

This section is about frequency dispersion for waves on a fluid layer forced by gravity, and according to linear theory.

Wave Propagation and Dispersion

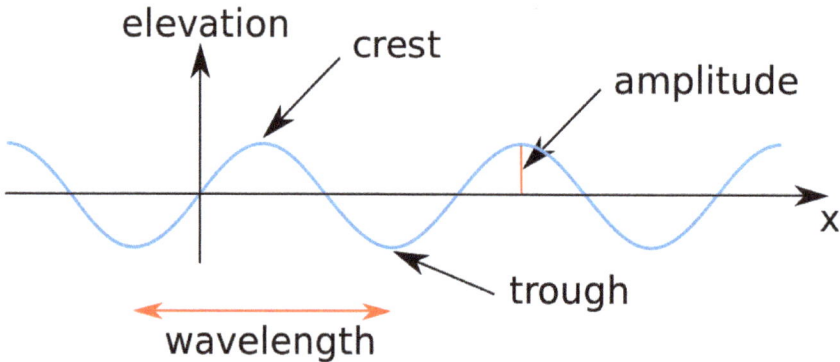

Sinusoidal wave

The simplest propagating wave of unchanging form is a sine wave. A sine wave with water surface elevation $\eta(x, t)$ is given by:

$$\eta(x,t) = a\sin\left(\theta(x,t)\right),$$

where a is the amplitude (in metres) and $\theta = \theta(x, t)$ is the phase function (in radians), depending on the horizontal position (x, in metres) and time (t, in seconds):

$$\theta = 2\pi\left(\frac{x}{\lambda} - \frac{t}{T}\right) = kx - \omega t, \quad \text{with} \quad k = \frac{2\pi}{\lambda} \quad \text{and} \quad \omega = \frac{2\pi}{T},$$

where:

- λ is the wavelength (in metres),

- T is the period (in seconds),

- k is the wavenumber (in radians per metre) and

- ω is the angular frequency (in radians per second).

Characteristic phases of a water wave are:

- the upward zero-crossing at $\theta = 0$,

- the wave crest at $\theta = \frac{1}{2}\pi$,

- the downward zero-crossing at $\theta = \pi$ and

- the wave trough at $\theta = 1\frac{1}{2}\pi$.

A certain phase repeats itself after an integer m multiple of 2π: $\sin(\theta) = \sin(\theta + m \cdot 2\pi)$.

Essential for water waves, and other wave phenomena in physics, is that free propagating waves of non-zero amplitude only exist when the angular frequency ω and wavenumber k (or equivalently the wavelength λ and period T) satisfy a functional relationship: the frequency dispersion relation

$$\omega^2 = \Omega^2(k).$$

The dispersion relation has two solutions: $\omega = +\Omega(k)$ and $\omega = -\Omega(k)$, corresponding to waves travelling in the positive or negative x–direction. The dispersion relation will in general depend on several other parameters in addition to the wavenumber k. For gravity waves, according to linear theory, these are the acceleration by gravity g and the water depth h. The dispersion relation for these waves is:

$$\omega^2 = gk\tanh(kh) \quad \text{or} \quad \lambda = \frac{g}{2\pi}T^2 \tanh\left(2\pi\frac{h}{\lambda}\right),$$

an implicit equation with tanh denoting the hyperbolic tangent function.

An initial wave phase $\theta = \theta_0$ propagates as a function of space and time. Its subsequent position is given by:

$$x = \frac{\lambda}{T}t + \frac{\lambda}{2\pi}\theta_0 = \frac{\omega}{k}t + \frac{\theta_0}{k}.$$

This shows that the phase moves with the velocity:

$$c_p = \frac{\lambda}{T} = \frac{\omega}{k} = \frac{\Omega(k)}{k},$$

which is called the phase velocity.

Phase Velocity

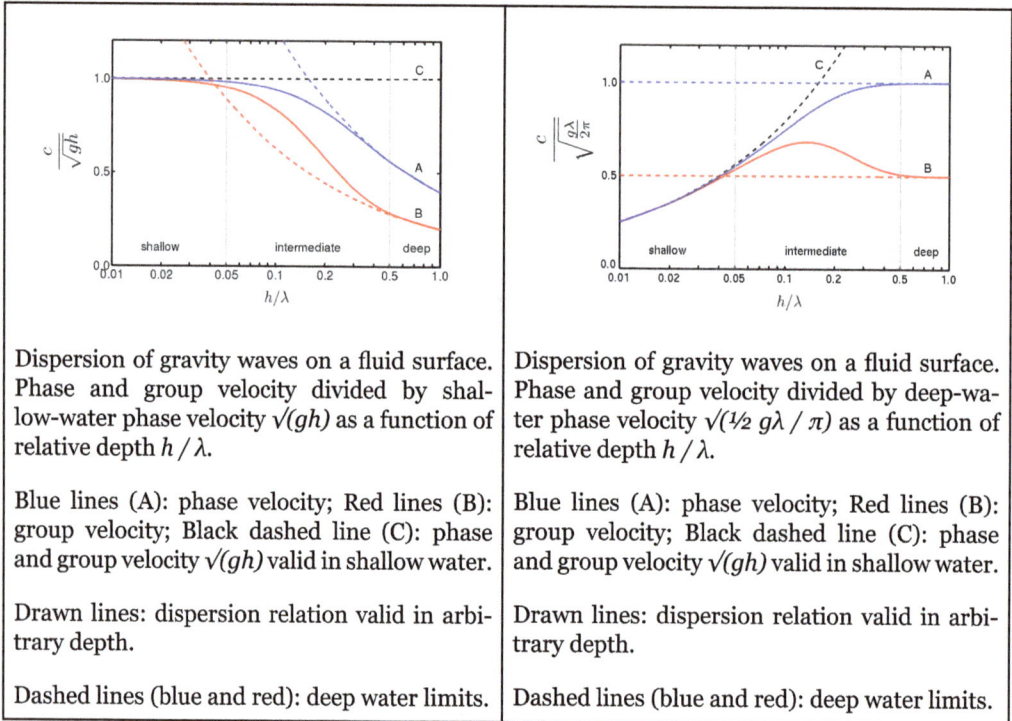

Dispersion of gravity waves on a fluid surface. Phase and group velocity divided by shallow-water phase velocity $\sqrt{(gh)}$ as a function of relative depth h / λ.	Dispersion of gravity waves on a fluid surface. Phase and group velocity divided by deep-water phase velocity $\sqrt{(\frac{1}{2} g\lambda / \pi)}$ as a function of relative depth h / λ.
Blue lines (A): phase velocity; Red lines (B): group velocity; Black dashed line (C): phase and group velocity $\sqrt{(gh)}$ valid in shallow water.	Blue lines (A): phase velocity; Red lines (B): group velocity; Black dashed line (C): phase and group velocity $\sqrt{(gh)}$ valid in shallow water.
Drawn lines: dispersion relation valid in arbitrary depth.	Drawn lines: dispersion relation valid in arbitrary depth.
Dashed lines (blue and red): deep water limits.	Dashed lines (blue and red): deep water limits.

A sinusoidal wave, of small surface-elevation amplitude and with a constant wavelength, propagates with the phase velocity, also called celerity or phase speed. While the phase velocity is a vector and has an associated direction, celerity or phase speed refer only to the magnitude of the phase velocity. According to linear theory for waves forced by gravity, the phase speed depends on the wavelength and the water depth. For a fixed water depth, long waves (with large wavelength) propagate faster than shorter waves.

In the left figure, it can be seen that shallow water waves, with wavelengths λ much larger than the water depth h, travel with the phase velocity

$$c_p = \sqrt{gh} \qquad \text{(shallow water)},$$

with g the acceleration by gravity and c_p the phase speed. Since this shallow-water phase speed is independent of the wavelength, shallow water waves do not have frequency dispersion.

Using another normalization for the same frequency dispersion relation, the figure on the right shows that in deep water, with water depth h larger than half the wavelength λ (so for $h/\lambda > 0.5$), the phase velocity c_p is independent of the water depth:

$$c_p = \sqrt{\frac{g\lambda}{2\pi}} = \frac{g}{2\pi}T \qquad \text{(deep water)},$$

with T the wave period (the reciprocal of the frequency f, $T=1/f$). So in deep water the phase speed increases with the wavelength, and with the period.

Since the phase speed satisfies $c_p = \lambda/T = \lambda f$, wavelength and period (or frequency) are related. For instance in deep water:

$$\lambda = \frac{g}{2\pi}T^2 \qquad \text{(deep water)}.$$

The dispersion characteristics for intermediate depth are given below.

Group Velocity

Frequency dispersion in bichromatic groups of gravity waves on the surface of deep water. The red dot moves with the phase velocity, and the green dots propagate with the group velocity.

More

In this deep-water case, the phase velocity is twice the group velocity. The red dot overtakes two green dots, when moving from the left to the right of the figure.

New waves seem to emerge at the back of a wave group, grow in amplitude until they are at the center of the group, and vanish at the wave group front.

For gravity surface-waves, the water particle velocities are much smaller than the phase velocity, in most cases.

Interference of two sinusoidal waves with slightly different wavelengths, but the same amplitude and propagation direction, results in a beat pattern, called a wave group. As can be seen in the animation, the group moves with a group velocity c_g different from the phase velocity c_p, due to frequency dispersion.

The group velocity is depicted by the red lines (marked B) in the two figures above. In shallow water, the group velocity is equal to the shallow-water phase velocity. This is because shallow water waves are not dispersive. In deep water, the group velocity is equal to half the phase velocity: $c_g = \frac{1}{2}\,c_p$.

The group velocity also turns out to be the energy transport velocity. This is the velocity with which the mean wave energy is transported horizontally in a narrow-band wave field.

In the case of a group velocity different from the phase velocity, a consequence is that the number of waves counted in a wave group is different when counted from a snapshot in space at a certain moment, from when counted in time from the measured surface elevation at a fixed position. Consider a wave group of length Λ_g and group duration of τ_g. The group velocity is:

$$c_g = \frac{\Lambda_g}{\tau_g}.$$

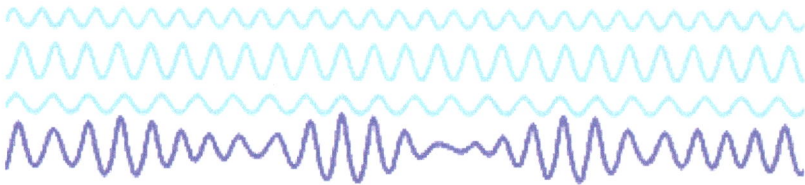

The number of waves per group as observed in space at a certain moment
(upper blue line), is different from the number of waves per group
seen in time at a fixed position (lower orange line), due to frequency dispersion

For the shown case, a bichromatic group of gravity waves on the surface of deep water, the group velocity is half the phase velocity. In this example, there are $5\frac{3}{4}$ waves between two wave group nodes in space, while there are $11\frac{1}{2}$ waves between two wave group nodes in time.

North Pacific storm waves as seen from the NOAA M/V Noble Star, Winter 1989

The number of waves in a wave group, measured in space at a certain moment is: Λ_g / λ. While measured at a fixed location in time, the number of waves in a group is: τ_g / T. So the ratio of the number of waves measured in space to those measured in time is:

$$\frac{\text{No. of waves in space}}{\text{No. of waves in time}} = \frac{\Lambda_g / \lambda}{\tau_g / T} = \frac{\Lambda_g}{\tau_g} \cdot \frac{T}{\lambda} = \frac{c_g}{c_p}.$$

So in deep water, with $c_g = \frac{1}{2}\, c_p$, a wave group has twice as many waves in time as it has in space.

The water surface elevation $\eta(x,t)$, as a function of horizontal position x and time t, for a bichromatic wave group of full modulation can be mathematically formulated as:

$$\eta = a\sin\left(k_1 x - \omega_1 t\right) + a\sin\left(k_2 x - \omega_2 t\right),$$

with:

- a the wave amplitude of each frequency component in metres,

- k_1 and k_2 the wave number of each wave component, in radians per metre, and

- ω_1 and ω_2 the angular frequency of each wave component, in radians per second.

Both ω_1 and k_1, as well as ω_2 and k_2, have to satisfy the dispersion relation:

$$\omega_1^2 = \Omega^2(k_1) \quad \text{and} \quad \omega_2^2 = \grave{U}^2(k_2).$$

Using trigonometric identities, the surface elevation is written as:

$$\eta = \left[2a\cos\left(\frac{k_1 - k_2}{2}x - \frac{\omega_1 - \omega_2}{2}t \right) \right] \cdot \sin\left(\frac{k_1 + k_2}{2}x - \frac{\omega_1 + \omega_2}{2}t \right).$$

The part between square brackets is the slowly varying amplitude of the group, with group wave number $\frac{1}{2}\,(k_1 - k_2)$ and group angular frequency $\frac{1}{2}\,(\omega_1 - \omega_2)$. As a result, the group velocity is, for the limit $k_1 \rightarrow k_2$:

$$c_g = \lim_{k_1 \rightarrow k_2} \frac{\omega_1 - \omega_2}{k_1 - k_2} = \lim_{k_1 \rightarrow k_2} \frac{\Omega(k_1) - \Omega(k_2)}{k_1 - k_2} = \frac{d\Omega(k)}{dk}.$$

Wave groups can only be discerned in case of a narrow-banded signal, with the wave-number difference $k_1 - k_2$ small compared to the mean wave number $\frac{1}{2}\,(k_1 + k_2)$.

Multi-component Wave Patterns

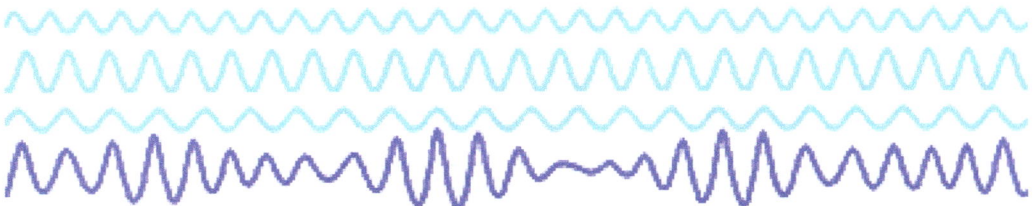

Frequency dispersion of surface gravity waves on deep water. The superposition (dark blue line) of three sinusoidal wave components (light blue lines) is shown

For the three components respectively 22 (bottom), 25 (middle) and 29 (top) wavelengths fit in a horizontal domain of 2,000 meter length. The component with the shortest wavelength (top) propagates slowest. The wave amplitudes of the components are respectively 1, 2 and 1 meter. The differences in wavelength and phase speed of the components results in a changing pattern of wave groups, due to amplification where the components are in phase, and reduction where they are in anti-phase.

The effect of frequency dispersion is that the waves travel as a function of wavelength, so that spatial and temporal phase properties of the propagating wave are constantly changing. For example, under the action of gravity, water waves with a longer wavelength travel faster than those with a shorter wavelength.

While two superimposed sinusoidal waves, called a bichromatic wave, have an envelope which travels unchanged, three or more sinusoidal wave components result in a changing pattern of the waves and their envelope. A sea state – that is: real waves on the sea or ocean – can be described as a superposition of many sinusoidal waves with different wavelengths, amplitudes, initial phases and propagation directions. Each of these components travels with its own phase velocity, in accordance with the dispersion relation. The statistics of such a surface can be described by its power spectrum.

Dispersion Relation

The full linear dispersion relation was first found by Pierre-Simon Laplace, although there were some errors in his solution for the linear wave problem. The complete theory for linear water waves, including dispersion, was derived by George Biddell Airy and published in about 1840. A similar equation was also found by Philip Kelland at around the same time (but making some mistakes in his derivation of the wave theory).

The shallow water (with small h / λ) limit, $\omega^2 = gh\, k^2$, was derived by Joseph Louis Lagrange.

Surface Tension Effects

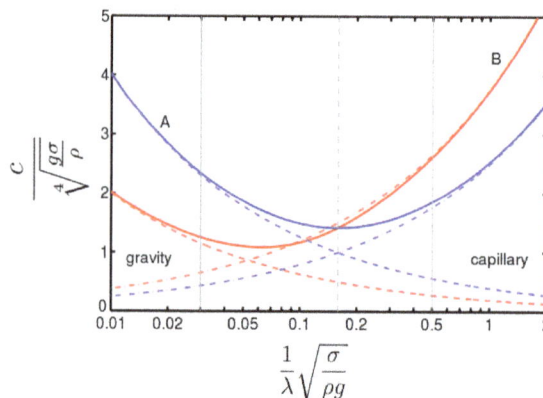

Dispersion of gravity-capillary waves on the surface of deep water. Phase and group velocity divided by $\sqrt[4]{g\sigma/\rho}$ as a function of inverse relative wavelength $\frac{1}{\lambda}\sqrt{\sigma/(\rho g)}$. ·

Blue lines (A): phase velocity, Red lines (B): group velocity.

Drawn lines: dispersion relation for gravity-capillary waves.

Dashed lines: dispersion relation for deep-water gravity waves.

Dash-dot lines: dispersion relation valid for deep-water capillary waves.

In case of gravity–capillary waves, where surface tension affects the waves, the dispersion relation becomes:

$$\omega^2 = \left(gk + \frac{\sigma}{\rho}k^3 \right)\tanh(kh),$$

with σ the surface tension (in N/m).

For a water–air interface (with σ = 0.074 N/m and ρ = 1000 kg/m³) the waves can be approximated as pure capillary waves – dominated by surface-tension effects – for wavelengths less than 0.4 cm (0.2 in). For wavelengths above 7 cm (3 in) the waves are to good approximation pure surface gravity waves with very little surface-tension effects.

Interfacial Waves

Wave motion on the interface between two layers of inviscid homogeneous fluids of different density, confined between horizontal rigid boundaries (at the top and bottom). The motion is forced by gravity. The upper layer has mean depth h' and density ρ', while the lower layer has mean depth h and density ρ. The wave amplitude is a, the wavelength is denoted by λ.

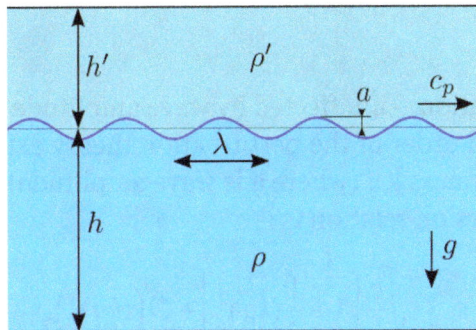

For two homogeneous layers of fluids, of mean thickness h below the interface and h' above – under the action of gravity and bounded above and below by horizontal rigid walls – the dispersion relationship $\omega^2 = \Omega^2(k)$ for gravity waves is provided by:

$$\Omega^2(k) = \frac{gk(\rho - \rho')}{\rho \coth(kh) + \rho' \coth(kh')},$$

where again ρ and ρ' are the densities below and above the interface, while coth is the hyperbolic cotangent function. For the case ρ' is zero this reduces to the dispersion relation of surface gravity waves on water of finite depth h.

When the depth of the two fluid layers becomes very large (h→∞, h'→∞), the hyperbolic cotangents in the above formula approaches the value of one. Then:

$$\Omega^2(k) = \frac{\rho - \rho'}{\rho + \rho'} gk.$$

Nonlinear Effects

Shallow Water

Amplitude dispersion effects appear for instance in the solitary wave: a single hump of water traveling with constant velocity in shallow water with a horizontal bed. Note that solitary waves are near-solitons, but not exactly – after the interaction of two (colliding or overtaking) solitary waves, they have changed a bit in amplitude and an oscillatory residual is left behind. The single soliton solution of the Korteweg–de Vries equation, of wave height H in water depth h far away from the wave crest, travels with the velocity:

$$c_p = c_g = \sqrt{g(h + H)}.$$

So for this nonlinear gravity wave it is the total water depth under the wave crest that determines the speed, with higher waves traveling faster than lower waves. Note that solitary wave solutions only exist for positive values of H, solitary gravity waves of depression do not exist.

Deep Water

The linear dispersion relation – unaffected by wave amplitude – is for nonlinear waves also correct at the second order of the perturbation theory expansion, with the orders in terms of the wave steepness k a (where a is wave amplitude). To the third order, and for deep water, the dispersion relation is

$$\omega^2 = gk\left[1 + (ka)^2\right], \quad \text{so} \quad c_p = \sqrt{\frac{g}{k}}\left[1 + \frac{1}{2}(ka)^2\right] + \mathcal{O}\left((ka)^4\right).$$

This implies that large waves travel faster than small ones of the same frequency. This is only noticeable when the wave steepness k a is large.

Waves on a Mean Current: Doppler Shift

Water waves on a mean flow (so a wave in a moving medium) experience a Doppler shift. Suppose the dispersion relation for a non-moving medium is:

$$\omega^2 = \Omega^2(k),$$

with k the wavenumber. Then for a medium with mean velocity vector V, the dispersion relationship with Doppler shift becomes:

$$\left(\omega - \mathbf{k}\cdot\mathbf{V}\right)^2 = \Omega^2(k),$$

where k is the wavenumber vector, related to k as: k = |k|. The dot product k•V is equal to: k•V = kV cos α, with V the length of the mean velocity vector V: V = |V|. And α the angle between the wave propagation direction and the mean flow direction. For waves and current in the same direction, k•V=kV.

Diffusion Equation

The diffusion equation is a partial differential equation. In physics, it describes the behavior of the collective motion of micro-particles in a material resulting from the random movement of each micro-particle. In mathematics, it is applicable in common to a subject relevant to the Markov process as well as in various other fields, such as the material sciences, information science, life science, social science, and so on. These subjects described by the diffusion equation are generally called Brown problems.

Statement

The equation is usually written as:

$$\frac{\partial \phi(\mathbf{r},t)}{\partial t} = \nabla\cdot\left[D(\phi,\mathbf{r})\,\nabla\phi(\mathbf{r},t)\right],$$

where $\phi(r, t)$ is the density of the diffusing material at location r and time t and D(ϕ, r) is the collective diffusion coefficient for density ϕ at location r; and ∇ represents the vector differential operator del. If the diffusion coefficient depends on the density then the equation is nonlinear, otherwise it is linear.

More generally, when D is a symmetric positive definite matrix, the equation describes anisotropic diffusion, which is written (for three dimensional diffusion) as:

$$\frac{\partial \phi(\mathbf{r},t)}{\partial t} = \sum_{i=1}^{3}\sum_{j=1}^{3}\frac{\partial}{\partial x_i}\left[D_{ij}(\phi,\mathbf{r})\frac{\partial \phi(\mathbf{r},t)}{\partial x_j}\right]$$

If D is constant, then the equation reduces to the following linear differential equation:

$$\frac{\partial \phi(\mathbf{r},t)}{\partial t} = D\nabla^2 \phi(\mathbf{r},t),$$

also called the heat equation.

History and Development

In mathematics, many phenomena in various science fields are expressed by using the well-known evolution equations. The diffusion equation is one of them and mathematically corresponds to the Markov process in relation to the normal distribution rule. In physics, the motion of diffusion particles corresponds to the well-known Brown motion satisfying the parabolic law. It is widely accepted that the Brown problem is a general term of investigating subjects in various science fields relevant to the Markov process, such as the material science, the information science, the life science, the social science, and so on. The extended diffusion equations are used for various sciences fields. In that case, they sometimes have a sink and source of their concerned elements, for example, such as a local equilibrium relation between native defects in silicon crystal in the material science or between predation and prey in the life science. We must then solve a system of diffusion equations. In the following, however, we discuss the fundamental diffusion equation of the so-called Fick's diffusion equation in relation to the material science.

In history, the heat equation proposed by Fourier in 1822 has been applied to investigating a temperature distribution in materials. In 1827, the so-called Brown motion was found, where the self-diffusion of water is visualized by pollen micro particles motion. Nevertheless, the Brown motion had not been recognized as a diffusion problem until the Einstein theory of Brown motion in 1905, although it was a typical diffusion problem. In 1855, Fick applied the heat equation to diffusion phenomena as it had been.

In accordance with the industrial requirement, the solid materials such as alloys, semiconductors, multilayer materials, and so on, have been widely fabricated. The heat treatment is indispensable for their fabrication processes then. The migration of particles in a solid material is caused by the heat treatment. In relation to the migration of their particles, the diffusion problems of various solid materials have been thus widely investigated, although the diffusion equation was mainly applied to problems of liquid material in an early stage after the Fick's proposition.

The Gauss's divergence theorem shows that the diffusion equation is valid in the solid, liquid and gas states in every material as a material conservation law, if there is no sink and source in the given diffusion system. It is also shows that the corresponding Fick's first law to the Fick's second law is mathematically incomplete without a constant diffusion flux relevant to the Brown motion in the localized space. The constant diffusion flux is indispensable for understanding the self-diffusion mechanism. The self-diffusion mechanism itself was not directly investigated, although it had been indirectly investigated by behavior of impurity diffusion in a pure material as shown in the Einstein's Brown theory and the Langevin equation.

We found that the diffusivity of diffusion equation depends generally on the concentration of diffusion particles. In that case, the diffusion equation becomes a nonlinear partial differential equation, and the mathematical solution is almost impossible, even if it is a case of the time and one dimension space coordinate . In accordance with the parabolic law , Boltzmann transformed the diffusion equation of , which is a nonlinear partial differential equation, into a nonlinear ordinary differential equation of in 1894. Since then, however, the Boltzmann transformation equation had not been mathematically solved until recently, although Matano empirically used it for analyzing interdiffusion problems in the metallurgy field.

Here, the analytical method of diffusion equation, which is extremely superior in calculation to the existing analytical method such as the integral transformation method of Fourier or Laplace and/or the variable separation method, was thus established in the parabolic space.

In 1947, Kirkendall found that an inert marker set at a point in a binary alloy moves from the initial sate point after the diffusion treatment. The phenomena are so called Kirkendall effect and it was considered that we cannot understand it from the existing theory of binary interdiffusion in those days. As a result, a new concept of intrinsic diffusion was then introduced for understanding the Kirkendall effect in the interdiffusion problems. Based on the intrinsic diffusion concept, Darken derived a relation between an interdiffusion coefficient and intrinsic diffusion coefficients in a binary interdiffusion in 1948. At present, however, it is revealed that the so-called Darken equation itself is mathematically wrong in the derivation process. Although the concentration of diffusion particles is a real quantity in physics, the temperature is a thermodynamic state quantity. As far as the shape of heat conduction material is unchangeable during a thermal treatment, the coordinate system of heat equation set in a material is a fixed one, since the coordinate system is not influenced by variations of the material internal structure. On the other hand, strictly speaking, the coordinate system of diffusion equation set in the diffusion field (solvent) is a moving one, since the origin of coordinate system is generally influenced by such variations.

When Fick proposed the diffusion equation, the Gauss divergence theorem had been already reported in 1840. Nevertheless, the problem of coordinate system of diffusion equation had not been mathematically investigated in accordance with the divergence theorem until recently. In general, however, it is indispensable for understanding the diffusion problems to discuss their coordinate systems, since it is, strictly speaking, considered that the diffusion particles, solvent particles and also the diffusion region space simultaneously move against the experimentation system in the diffusion region outside.

Recently, the diffusion equation was thus mathematically investigated in accordance with the divergence theorem and the coordinate transformation theory. As a result, the diffusion flux should be determined by taking account of the concerned coordinate system of diffusion equation. Using the corresponding diffusion flux to the concerned coordinate system of diffusion equation for interdiffusion, one way diffusion, impurity diffusion and self-diffusion, we found that they are uniformly discussed and the foundation of diffusion problems is included in interdiffusion problems. The interdiffusion theory of an elements system applicable to every material was thus reasonably established. In the analysis of interdiffusion problems, the only difference between a binary system and an N elements system is whether the solvent material is one element or elements.

The coordinate transformation theory reveals that the Kirkendall effect is caused by a shift between the coordinate systems of diffusion equation like the Doppler effect relevant to a wave equation is caused by a shift between the fixed coordinate system and the moving one. Further, it was also found that the concept of intrinsic diffusion is an illusion in the diffusion history. All physical information in the given diffusion system is incorporated into the diffusivity. If we can know a diffusivity behavior in the given diffusion equation, the mathematical solution and/or numerical one at least is possible. In the diffusion problems, it is thus extremely dominant to know the diffusivity behavior. The diffusivity is defined by an interaction between a diffusion particle and the diffusion field near the diffusion particle itself. This indicates that the diffusivity should be essentially investigated in the quantum mechanics, since the behavior of a micro particle should be investigated by analyzing the Schrodinger equation.

From applying the diffusion equation to a problem of diffusion elementary process, the equation was reasonably derived. It was revealed that the diffusivity corresponds to the angular momentum operator in the quantum mechanics. As a result, the universal expression of diffusivity, which is applicable to every material in an arbitrary thermodynamic state, was obtained as one with the proportionality constant composed of the product of Planck constant and Avogadro constant . It was also found that the well-known material wave relation proposed by de Broglie in 1923, which is the most fundamental one in materials science, is obtained from a relation between the given diffusivity expressions. This gives evidence for the theory discussed here.

Derivation

The diffusion equation can be trivially derived from the continuity equation, which states that a change in density in any part of the system is due to inflow and outflow of material into and out of that part of the system. Effectively, no material is created or destroyed:

$$\frac{\partial \phi}{\partial t} + \nabla \cdot \mathbf{j} = 0,$$

where j is the flux of the diffusing material. The diffusion equation can be obtained easily from this when combined with the phenomenological Fick's first law, which states that the flux of the diffusing material in any part of the system is proportional to the local density gradient:

$$\mathbf{j} = -D(\phi,\mathbf{r})\nabla \phi(\mathbf{r},t).$$

If drift must be taken into account, the Smoluchowski equation provides an appropriate generalization.

Discretization

The diffusion equation is continuous in both space and time. One may discretize space, time, or both space and time, which arise in application. Discretizing time alone just corresponds to taking time slices of the continuous system, and no new phenomena arise. In discretizing space alone, the Green's function becomes the discrete Gaussian kernel, rather than the continuous Gaussian kernel. In discretizing both time and space, one obtains the random walk.

Discretization (Image)

The product rule is used to rewrite the anisotropic tensor diffusion equation, in standard discretization schemes. Because direct discretization of the diffusion equation with only first order spatial central differences leads to checkerboard artifacts. The rewritten diffusion equation used in image filtering:

$$\frac{\partial \phi(\mathbf{r},t)}{\partial t} = \nabla \cdot \left[D(\phi,\mathbf{r}) \right] \nabla \phi(\mathbf{r},t) + \mathrm{tr}\left[D(\phi,\mathbf{r})\left(\nabla \nabla^T \phi(\mathbf{r},t) \right) \right]$$

where "tr" denotes the trace of the 2nd rank tensor, and superscript "T" denotes transpose, in which in image filtering D(ϕ, r) are symmetric matrices constructed from the eigenvectors of the image structure tensors . The spatial derivatives can then be approximated by two first order and a second order central finite differences. The resulting diffusion algorithm can be written as an image convolution with a varying kernel (stencil) of size 3×3 in 2D and $3 \times 3 \times 3$ in 3D.

Analytical Solution of Diffusion Equation

In this lesson, we will obtain the analytical solution of the one dimensional diffusion equation for instantaneous release of mass at a particular point. The one dimensional diffusion equation as derived earlier is,

$$\frac{\partial C}{\partial t} = D\frac{\partial^2 C}{\partial x^2}$$

The solution of the diffusion equation gives the spatial distribution of concentration at different time. Let us consider the figure shown below.

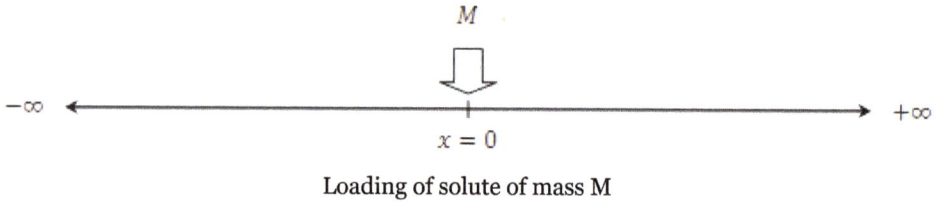

Loading of solute of mass M

Solute mass of M is instantaneously loaded at x = 0 and t = 0. The initial and boundary condition of the problem can be written as,

Initial condition,

$$C(x = 0, t = 0) = M\delta(x)$$

Where δ is the Kronecker delta. $\delta(x) = 0$ for $x \neq 0$ and $\delta(x) = 1$ for $x = 0$.

Boundary conditions

When $x \to \pm\infty$ $C \to$

When $x \to \pm\infty$ $\partial C / \partial x \to 0$

The basic technique use for solving this type of problem is to convert the partial differential equation to an ordinary differential equation using a dimensionless similarity variable. The dimensionless similarity variable can be obtained using dimensional analysis. The variables that control the distribution of concentration in a diffusion medium are M, D, x, and t. The general form of expression for C may be written as,

C = φ(M, D, x, t)

The dimensions of the variable involved can be tabulated as follows.

	C	M	D	x	t
Mass (M)	1	1	0	0	0
Length (L)	-1	0	2	1	0
Time (T)	0	0	-1	0	1

The number of variable here is 5 and the number of basic dimension is 3. Therefore the number of dimensionless group is 5-3=2. Thus, there will be three repeating variables to non-dimensionalize the group. It is useful to relate the concentration with the M, D, and t. Thus it will be reasonable to consider M, D, and t as the repeating variables. Then C and x will be the independent variable. The required solution may be written as,

$$\pi_1 = \varphi(\pi_2)$$

Where

$$\pi_1 = \frac{C}{M^p D^q t^r}$$

$$\pi_2 = \frac{x}{M^{p_1} D^{q_1} t^{r_1}}$$

Now, equating the powers M, D, and t of equation above to make the π_1 dimensionless, we have

$$
\left.
\begin{array}{l}
1p + 0q + 0r = 1 \\
0p + 2q + 0r = -1 \\
0p - 1q + 1r = 0
\end{array}
\right\}
$$

Solving, we have $p = 1,\ q = -\dfrac{1}{2}$ and $,r = -\dfrac{1}{2}$

Thus

$$\pi_1 = \frac{C}{\dfrac{M}{\sqrt{Dt}}}$$

Similarly equating the powers of M, D, and t of equation above to make π_2 dimensionless, we have

$$
\left.
\begin{array}{l}
1p_1 + 0q_1 + 0r_1 = 0 \\
0p_1 + 2q_1 + 0r_1 = 1 \\
0p_1 + 1q_1 + 1r_1 = 0
\end{array}
\right\}
$$

Solving, swe have $p = 0, q = \dfrac{1}{2}$ and $r = \dfrac{1}{2}$

$$\pi_2 = \frac{x}{\sqrt{Dt}}$$

Putting in equation above, we have

$$\frac{C}{\dfrac{M}{\sqrt{Dt}}} = \varphi\left(\frac{x}{\sqrt{Dt}}\right)$$

$$C = \frac{M}{\sqrt{Dt}}\varphi\left(\frac{x}{\sqrt{Dt}}\right)$$

$$C = \frac{M}{\sqrt{Dt}}\varphi\left(\frac{x}{\sqrt{Dt}}\right)$$

Putting $\quad \mu = \frac{M}{\sqrt{Dt}} \ and \ \eta\frac{x}{\sqrt{Dt}}$

$$C = \mu\varphi(\eta)$$

Now differentiating equation above with respect to time, we have

$$\frac{\partial C}{\partial t} = \frac{\partial}{\partial t}\left(\mu\varphi(\eta)\right)$$

From next step onward we will write φ in place of φ (η)

$$\Rightarrow \frac{\partial C}{\partial t} = \mu\frac{\partial \varphi}{\partial t} + \varphi\frac{\partial \mu}{\partial t}$$

$$\Rightarrow \frac{\partial C}{\partial t} = \mu\frac{\partial \varphi}{\partial t}\frac{\partial \eta}{\partial t} + \varphi\frac{\partial \mu}{\partial t}$$

$$\Rightarrow \frac{\partial C}{\partial t} = \mu\frac{\partial \varphi}{\partial \eta}\left(\frac{x}{\sqrt{D}}\right)\frac{\partial}{\partial t}\left(t^{-\frac{1}{2}}\right) + \varphi\left(\frac{M}{\sqrt{D}}\right)\frac{\partial}{\partial t}\left(t^{-\frac{1}{2}}\right)$$

$$\Rightarrow \frac{\partial C}{\partial t} = \frac{M}{\sqrt{D}}\frac{\partial \varphi}{\partial \eta}\left(\frac{x}{\sqrt{D}}\right)\frac{\partial}{\partial t}\left(t^{-\frac{1}{2}}\right) + \varphi\left(\frac{M}{\sqrt{D}}\right)\frac{\partial}{\partial t}\left(t^{-\frac{1}{2}}\right)$$

$$\Rightarrow \frac{\partial C}{\partial t} = \frac{M}{\sqrt{D}}\frac{\partial}{\partial t}\left(t^{-\frac{1}{2}}\right)\left[\frac{\partial \varphi}{\partial \eta}\frac{x}{\sqrt{D}} + \varphi\right]$$

$$\Rightarrow \frac{\partial C}{\partial t} = \frac{M}{\sqrt{D}}\left(-\frac{1}{2}t^{-\frac{1}{2}}\right)\left[\eta\frac{\partial \varphi}{\partial \eta} + \varphi\right]$$

As φ is a function of η only, we can write $\dfrac{\partial \varphi}{\partial \eta} = \dfrac{d\varphi}{d\eta}$, Thus

$$\Rightarrow \frac{\partial C}{\partial t} = -\frac{M}{2t\sqrt{Dt}}\left[\eta\frac{d\varphi}{d\eta} + \varphi\right]$$

Similarly, differentiating equation above with respect to x, we have

$$\frac{\partial C}{\partial x} = \frac{\partial}{\partial x}(\mu\varphi)$$

$$\Rightarrow \frac{\partial C}{\partial x} = \mu\frac{\partial\varphi}{\partial x}$$

$$\Rightarrow \frac{\partial C}{\partial x} = \mu\frac{\partial\varphi}{\partial\eta}\frac{\partial\eta}{\partial x}$$

$$\Rightarrow \frac{\partial C}{\partial x} = \mu\frac{\partial\varphi}{\partial\eta}\frac{\partial}{\partial x}\left(\frac{x}{\sqrt{D}}\right)$$

$$\Rightarrow \frac{\partial C}{\partial x} = \frac{M}{\sqrt{Dt}}\frac{\partial\varphi}{\partial\eta}\frac{1}{\sqrt{Dt}}$$

$$\Rightarrow \frac{\partial C}{\partial x} = \frac{M}{Dt}\frac{\partial\varphi}{\partial\eta}$$

Now,

$$\Rightarrow \frac{\partial^2 C}{\partial x^2} = \frac{\partial}{\partial x}\left(\frac{\partial C}{\partial x}\right)$$

Putting equation above, we have

$$\Rightarrow \frac{\partial^2 C}{\partial x^2} = \frac{\partial}{\partial x}\left(\frac{M}{Dt}\frac{\partial\varphi}{\partial\eta}\right)$$

$$\Rightarrow \frac{\partial^2 C}{\partial x^2} = \left(\frac{M}{Dt}\right)\frac{\partial}{\partial\eta}\left(\frac{\partial\varphi}{\partial\eta}\right)\frac{\partial\eta}{\partial x}$$

$$\Rightarrow \frac{\partial^2 C}{\partial x^2} = \left(\frac{M}{Dt}\right)\frac{\partial^2\varphi}{\partial\eta^2}\left(\frac{1}{\sqrt{Dt}}\right)$$

$$\Rightarrow \frac{\partial^2 C}{\partial x^2} = \frac{M}{(Dt)^{3/2}}\frac{d^2\varphi}{d\eta^2}$$

Putting the value of $\frac{\partial C}{\partial t}$ and $\frac{\partial^2 C}{\partial x^2}$ the diffusion equation, we have

$$-\frac{M}{2t\sqrt{Dt}}\left[\eta\frac{d\varphi}{d\eta}+\varphi\right]=D\frac{M}{(Dt)^{3/2}}\frac{d^2\varphi}{d\eta^2}$$

$$\Rightarrow \frac{d^2\varphi}{d\eta^2}+\frac{1}{2}\eta\frac{d\varphi}{d\eta}+\frac{1}{2}\varphi=0$$

The equation above is an ordinary differential equation of φ with respect to η.

Integrating, we have

$$\frac{d\varphi}{d\eta}+\frac{1}{2}\eta\varphi=\alpha$$

where, α is integration constant.

Now considering the boundary condition

When $x \to \pm\infty$ $\eta \to \pm\infty$ $c \to 0$

When $x \to \pm\infty$ $\eta \to \pm\infty$ $\partial C / \partial x \to 0$

Thus,

When $\eta \to \pm\infty$ $\alpha \to 0$

Therefore, the equation above becomes

$$\frac{d\varphi}{d\eta}+\frac{1}{2}\eta\varphi=0$$

$$\Rightarrow \frac{d\varphi}{d\eta}=-\frac{1}{2}\eta\varphi$$

$$\Rightarrow \frac{d\varphi}{\varphi}=-\frac{1}{2}\eta d\eta$$

Integrating,

$$\Rightarrow \int \frac{d\varphi}{\varphi}=-\frac{1}{2}\int \eta d\eta$$

$$\Rightarrow ln(\varphi) = -\frac{1}{2}\frac{\eta^2}{2} + \beta$$

$$\Rightarrow \varphi = e^{\left(-\frac{\eta^2}{4} + \beta\right)}$$

$$\Rightarrow \varphi = e^{\beta} e^{\left(-\frac{\eta^2}{4}\right)}$$

Taking $\gamma = e^{\beta}$

$$\Rightarrow \varphi = \gamma e^{-\left(\frac{\eta^2}{4}\right)}$$

Putting the above equations, we have

$$C(x,t) = \mu\gamma e^{-\left(\frac{\eta^2}{4}\right)}$$

$$\Rightarrow C(x,t) = \gamma\left(\frac{M}{\sqrt{Dt}}\right)e^{-\left(\frac{x^2}{4Dt}\right)}$$

Now for calculating the value of γ, integrate the equation above between $-\infty$ to $+\infty$

$$\int_{-\infty}^{+\infty} C(x,t)\,dx = \gamma\left(\frac{M}{\sqrt{Dt}}\right)\int_{-\infty}^{+\infty} e^{-\left(\frac{x^2}{4Dt}\right)}dx$$

The term $\int_{-\infty}^{+\infty} C(x,t)\,dx$ is the total injected mass at t = 0, which is equal to M.
Therefore,

$$M = \gamma\left(\frac{M}{\sqrt{Dt}}\right)\int_{-\infty}^{+\infty} e^{-\left(\frac{\eta^2}{4}\right)}\sqrt{Dt}\,d\eta$$

$$\Rightarrow M = \gamma M\int_{-\infty}^{+\infty} e^{-\left(\frac{\eta^2}{4}\right)}d\eta$$

$$\Rightarrow 1 = \gamma \sqrt{\frac{\pi}{\frac{1}{4}}} = \gamma \sqrt{2\pi} \qquad \left[As \int_{-\infty}^{+\infty} e^{-ax^2} dx = \sqrt{\frac{\pi}{a}} \right]$$

$$\Rightarrow \gamma = \frac{1}{\sqrt{2\pi}}$$

Thus the equation above becomes

$$\Rightarrow C(x,t) = \frac{1}{\sqrt{2\pi}} \left(\frac{M}{\sqrt{Dt}} \right) e^{-\left(\frac{x^2}{4Dt} \right)}$$

$$\Rightarrow C(x,t) = \left(\frac{M}{\sqrt{4\pi Dt}} \right) e^{-\left(\frac{x^2}{4Dt} \right)}$$

This is the solution of the one dimensional diffusion equation for the case of instantaneous loading at x= 0 and at t = 0. If the release of mass is at any arbitrary location x = x_1 and at arbitrary time t = t_1 as shown in the figure below.

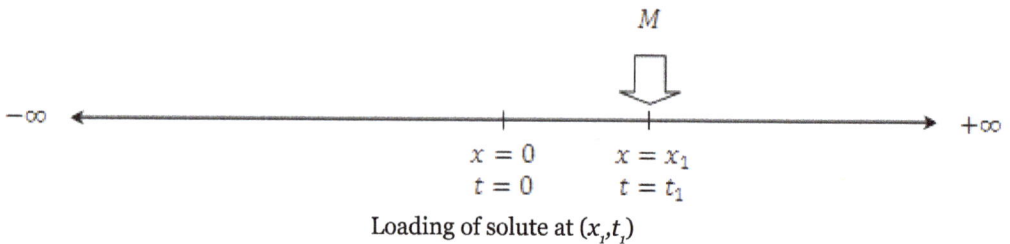

Loading of solute at (x_1, t_1)

Then,

$$\left. \begin{array}{l} t' = t - t_1 \\ x' = x - x_1 \end{array} \right\}$$

Thus the equation above becomes

$$\Rightarrow C(x,t) = \left(\frac{M}{\sqrt{4\pi D(t - t_1)}} \right) e^{-\left(\frac{(x - x_1)^2}{4Dt} \right)}$$

Figure below shows the spatial and temporal distribution of solute spread by the process of diffusion due to the instantaneous loading of solute at x = 0 and at time t = 0. This solution is obtained using equation above.

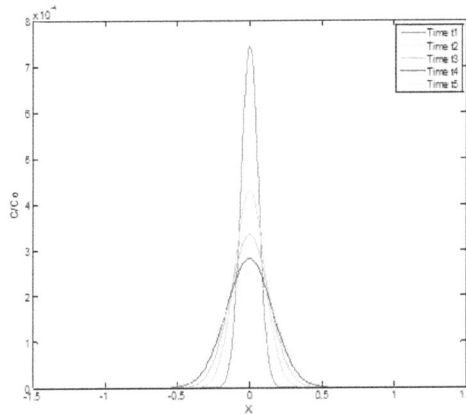

Spatial and temporal distribution of solute due to instantaneous loading at $(x = 0, t = 0)$

Figure below shows the spatial and temporal distribution of solute spread by the process of diffusion due to the instantaneous loading of solute at $x_1 = 0.5$ and at $t = 0$. This solution is obtained using equation above.

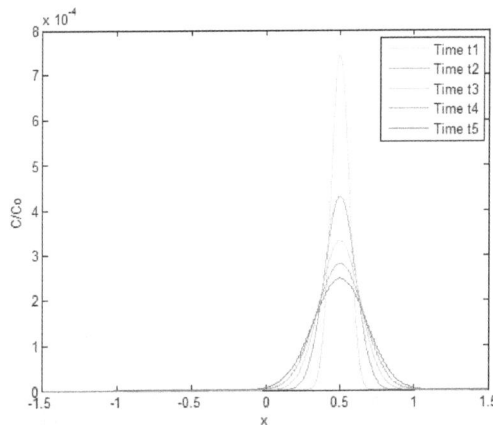

Spatial and temporal distribution of solute due to instantaneous loading at $(x_1 = 0.5, t = 0)$

The equation above can also be written as

$$\frac{C(x,t)}{M} = \left(\frac{1}{\sqrt{4\pi Dt}}\right) e^{-\left(\frac{(x-x_1)^2}{4Dt}\right)}$$

We know that the Gaussian distribution is

$$f(x) = \left(\frac{1}{\sigma_n \sqrt{2\pi}}\right) e^{-\frac{1}{2}\left(\frac{x-\mu}{\sigma_n}\right)^2}$$

Where σ_n is the standard deviation and μ is the mean.

The equations above are equal if

$$\sigma_n^2 = 2Dt$$

$$\Rightarrow D = \frac{\sigma_n^2}{2t}$$

The equation above can be used to calculate the value of coefficient of diffusion using experimental data.

Solution of Two-dimensional Diffusion Equation

In this chapter, we will obtain analytical solution of 2D and 3D diffusion equations using separation of variable technique.

The 2D form of the diffusion equation can be written as:

$$D_x \frac{\partial^2 C}{\partial x^2} + D_y \frac{\partial^2 C}{\partial y^2} = \frac{\partial C}{\partial t}$$

Consider the 2D unbounded domain (Figure below). The solution of the diffusion equation will give us the spatial and temporal distribution concentration in the unbounded domain.

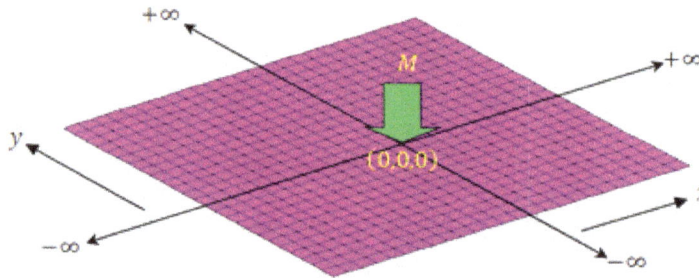

Loading of solute of mass M in a 2D unbounded system

For the unbounded domain, the initial and boundary condition are,

Initial condition

$$C (x = 0, y = 0, t = 0) = M\delta(x,y)$$

Where δ is the Kronecker delta. $\delta(x, y) = 0$ for x, y $\neq$ 0 and $\delta(x) = 1$ for x, y = 0.

Boundary conditions

When $x \rightarrow \pm\infty$ $\partial C / \partial x, C \rightarrow 0$

When $y \rightarrow \pm\infty$ $\partial C / \partial y, C \rightarrow 0$

The diffusion equation for the initial and boundary condition can be solved using separation of variable method. Let us consider that

$$C(x, y, t) = C_1(x, t) C_2(y, t)$$

Now,

$$\frac{\partial^2 C}{\partial x^2} = \frac{\partial}{\partial x}\left(\frac{\partial}{\partial x}(C_1 C_2)\right)$$

$$\Rightarrow \frac{\partial^2 C}{\partial x^2} = \frac{\partial}{\partial x}\left(C_2 \frac{\partial C_1}{\partial x}\right)$$

$$\Rightarrow \frac{\partial^2 C}{\partial x^2} = C_2 \frac{\partial^2 C_1}{\partial x^2}$$

Similarly,

$$\Rightarrow \frac{\partial^2 C}{\partial y^2} = C_1 \frac{\partial^2 C_2}{\partial y^2}$$

Again,

$$\frac{\partial C}{\partial t} = C_1 \frac{\partial C_2}{\partial t} + C_2 \frac{\partial C_2}{\partial t}$$

Putting together the equations above, we have

$$D_x C_2 \frac{\partial^2 C_1}{\partial x^2} + D_y C_1 \frac{\partial^2 C_2}{\partial y^2} = C_1 \frac{\partial C_2}{\partial t} + C_2 \frac{\partial C_1}{\partial t}$$

$$\Rightarrow C_1 \left(\frac{\partial C_2}{\partial t} - D_y \frac{\partial^2 C_2}{\partial y^2}\right) + C_2 \left(\frac{\partial C_1}{\partial t} - D_x \frac{\partial^2 C_1}{\partial x^2}\right) = 0$$

Since $C_1 \neq 0$ and $C_2 \neq 0$

Thus,

$$\frac{\partial C_1}{\partial t} - D_x \frac{\partial^2 C_1}{\partial x^2} = 0$$

$$\frac{\partial C_2}{\partial t} - D_y \frac{\partial^2 C_2}{\partial y^2} = 0$$

We have obtained the solution of the equations above. The solutions are,

$$C_1(x,t) = \left(\frac{\alpha}{\sqrt{4\pi D_x t}} \right) e^{-\left(\frac{x^2}{4D_x t} \right)}$$

$$C_2(y,t) = \left(\frac{\beta}{\sqrt{4\pi D_y t}} \right) e^{-\left(\frac{y^2}{4D_y t} \right)}$$

Putting together the equations above, we have

$$C(x,y,t) = \left(\frac{\alpha}{\sqrt{4\pi D_x t}} \right) e^{-\left(\frac{x^2}{4D_x t} \right)} \times \left(\frac{\beta}{\sqrt{4\pi D_y t}} \right) e^{-\left(\frac{y^2}{4D_y t} \right)}$$

$$\Rightarrow C(x,y,t) = \left(\frac{\alpha\beta}{4\pi t\sqrt{D_x D_y}} \right) e^{-\left(\frac{x^2}{4D_x t} \right)} \times e^{-\left(\frac{y^2}{4D_y t} \right)}$$

$$\Rightarrow C(x,y,t) = \left(\frac{\gamma}{4\pi t\sqrt{D_x D_y}} \right) e^{-\left(\frac{x^2}{4D_x t} + \frac{y^2}{4D_y t} \right)}$$

Where, $\gamma = \alpha\beta$

The double integration of C (x, y, t) from $-\infty$ to $+\infty$ is equal to the total mass injected at t = 0 at (0, 0).

$$\int_{-\infty}^{+\infty} \int_{-\infty}^{+\infty} C(x,y,t)\, dx\, dy = M$$

Thus

$$\int_{-\infty}^{+\infty} \int_{-\infty}^{+\infty} \left(\frac{\gamma}{4\pi t\sqrt{D_x D_y}} \right) e^{-\left(\frac{x^2}{4D_x t} + \frac{y^2}{4D_y t} \right)}\, dx\, dy = M$$

$$\Rightarrow \left(\frac{\gamma}{4\pi t\sqrt{D_x D_y}} \right) \int_{-\infty}^{+\infty} \int_{-\infty}^{+\infty} e^{-\left(\frac{x^2}{4D_x t} + \frac{y^2}{4D_y t} \right)}\, dx\, dy = M$$

Considering, $a = \dfrac{1}{4D_x t}$ and $= \dfrac{1}{4D_y t}$, we have

$$\left(\frac{\gamma}{4\pi t\sqrt{D_x D_y}}\right) \int_{-\infty}^{+\infty} \int_{-\infty}^{+\infty} e^{-ax^2} e^{-by^2} \, dxdy = M$$

$$\Rightarrow \left(\frac{\gamma}{4\pi t\sqrt{D_x D_y}}\right) \int_{-\infty}^{+\infty} e^{-by^2} \left(\int_{-\infty}^{+\infty} e^{-ax^2} \, dx\right) dy = M$$

$$\Rightarrow \left(\frac{\gamma}{4\pi t\sqrt{D_x D_y}}\right) \int_{-\infty}^{+\infty} e^{-by^2} \sqrt{\frac{\pi}{a}} \, dy = M \qquad \left[As \int_{-\infty}^{+\infty} e^{-ax^2} \, dx \sqrt{\frac{\pi}{a}} \right]$$

$$\Rightarrow \left(\frac{\gamma}{4\pi t\sqrt{D_x D_y}}\right) \sqrt{\frac{\pi}{a}} \left(\int_{-\infty}^{+\infty} e^{-by^2} \, dy\right) = M$$

$$\Rightarrow \left(\frac{\gamma}{4\pi t\sqrt{D_x D_y}}\right) \sqrt{\frac{\pi}{a}} \sqrt{\frac{\pi}{a}} = M$$

Putting the value of a and b in equation above, we have

$$\left(\frac{\gamma}{4\pi t\sqrt{D_x D_y}}\right) \sqrt{4D_x t\pi} \sqrt{4D_y t\pi} = M$$

$$\Rightarrow \left(\frac{\gamma}{4\pi t\sqrt{D_x D_y}}\right) 4\pi t\sqrt{D_x D_y} = M$$

$$\Rightarrow \gamma = M$$

Thus,

$$C(x, y, t) = \left(\frac{M}{4\pi t\sqrt{D_x D_y}}\right) e^{-\left(\frac{x^2}{4D_x t} + \frac{y^2}{4D_y t}\right)}$$

$$C(x, y, t) = \left(\frac{M}{4\pi t\sqrt{D_x D_y}}\right) e^{-\frac{1}{4t}\left(\frac{x^2}{4D_x} + \frac{y^2}{4D_y}\right)}$$

This is the solution of the two dimensional diffusion equation for the case of instantaneous loading at x = 0, y = 0 and at t = 0. If the release of mass is at any arbitrary location = x_1, y = y_1 and at arbitrary time t = t_1 as shown in the figure below.

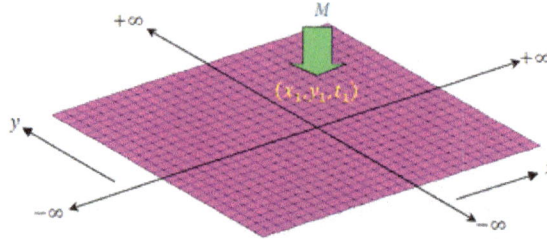

Loading of solute of mass M at (x_1, y_1, t_1) in a 2D unbounded domain

Then,

$$\left. \begin{aligned} x' &= x - x_1 \\ y' &= y - y_1 \\ t' &= t - t_1 \end{aligned} \right\}$$

Thus the equation above becomes

$$c(x, y, t) = \left(\frac{M}{4\pi(t - t_1)\sqrt{D_x D_y}} \right) e^{-\frac{1}{(t-t_1)}\left(\frac{(x-x_1)^2}{D_x} + \frac{(y-y_1)^2}{D_y} \right)}$$

Figure below shows the spatial and temporal distribution of solute spread by the process of diffusion due to the instantaneous loading of solute at x = 0, y = 0 and at time t = 0. This solution is obtained using equation above.

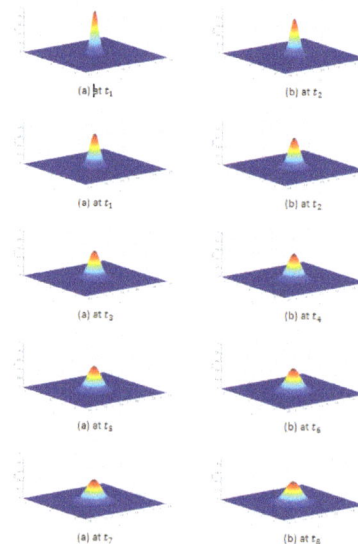

(a) at t_1 (b) at t_2

(a) at t_1 (b) at t_2

(a) at t_2 (b) at t_4

(a) at t_5 (b) at t_6

(a) at t_7 (b) at t_8

Spatial and temporal distribution of solute due to instantaneous loading at (x = 0, y = 0, t = 0)

Solution of Three-dimensional Diffusion Equation

The separation of variable technique can also be applied to obtain analytical solution of 3D diffusion equation. The 3D form of the diffusion equation can be written as

$$D_x \frac{\partial^2 C}{\partial x^2} + D_y \frac{\partial^2 C}{\partial y^2} + D_z \frac{\partial^2 C}{\partial z^2} = \frac{\partial C}{\partial t}$$

Consider the 3D unbounded domain shown in the figure below. The solution of the diffusion equation will give us the spatial and temporal distribution concentration in the unbounded domain.

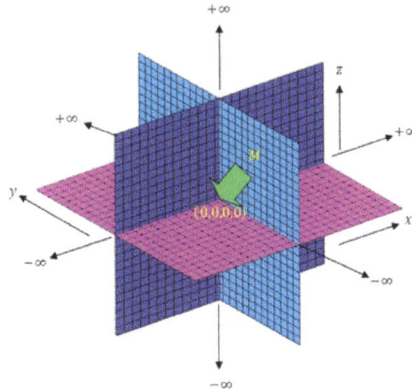

Loading of solute of mass M in a 3D unbounded system

For the unbounded domain, the initial and boundary condition are,

Initial condition

$$C(x= 0, y= 0, y=0, t= 0) = M\delta(x, y, z)$$

Where δ is the Kronecker delta $\delta(x, y, z) = 0$ *for* $x, y, z \neq 0$ *and* $\delta(x) = 1$ *for* $x, y, z = 0$.

Boundary conditions

When $x \rightarrow \pm\infty$ $\qquad\qquad$ $\partial C / \partial x, C \rightarrow 0$

When $y \rightarrow \pm\infty$ $\qquad\qquad$ $\partial C / \partial y, C \rightarrow 0$

When $z \rightarrow \pm\infty$ $\qquad\qquad$ $\partial C / \partial z, C \rightarrow 0$

Like the 2D solution of the diffusion equation, the 3D diffusion equation above, for the initial and boundary condition can also be solved using separation of variable method. Let us consider that

$$C(x, y, t) = C_1(x, t) C_2(y, t) C_3(z, t)$$

Now

$$\frac{\partial^2 C}{\partial x^2} = \frac{\partial}{\partial x}\left(\frac{\partial}{\partial x}(C_1 C_2 C_3)\right)$$

$$\Rightarrow \frac{\partial^2 C}{\partial x^2} = \frac{\partial}{\partial x}\left(C_2 C_3 \frac{\partial C_1}{\partial x}\right)$$

$$\Rightarrow \frac{\partial^2 C}{\partial x^2} = C_2 C_3 \frac{\partial^2 C_1}{\partial x^2}$$

Similarly

$$\Rightarrow \frac{\partial^2 C}{\partial y^2} = C_3 C_1 \frac{\partial^2 C_2}{\partial y^2}$$

$$\Rightarrow \frac{\partial^2 C}{\partial z^2} = C_1 C_2 \frac{\partial^2 C_2}{\partial z^2}$$

Again

$$\frac{\partial C}{\partial t} = C_2 C_3 \frac{\partial C_1}{\partial t} + C_3 C_1 \frac{\partial C_2}{\partial t} + C_1 C_2 \frac{\partial C_3}{\partial t}$$

Putting together the equations above, we have

$$D_x C_2 C_3 \frac{\partial^2 C_1}{\partial x^2} + D_y C_3 C_1 \frac{\partial^2 C_2}{\partial y^2} + D_z C_1 C_2 \frac{\partial^2 C_3}{\partial t} = C_2 C_3 \frac{\partial C_1}{\partial t} + C_3 C_1 \frac{\partial C_2}{\partial t} + C_1 C_2 \frac{\partial C_3}{\partial t}$$

$$\Rightarrow C_2 C_3 \left(\frac{\partial C_1}{\partial t} - D_x \frac{\partial^2 C_1}{\partial x^2}\right) + C_3 C_1 \left(\frac{\partial C_2}{\partial t} - D_y \frac{\partial^2 C_2}{\partial y^2}\right) + C_1 C_2 \left(\frac{\partial C_3}{\partial t} - D_z \frac{\partial^2 C_3}{\partial z^2}\right) = 0$$

Since $C_2 C_3 \neq 0, C_3 C_1 \neq 0$ *and* $C_1 C_2 \neq 0$

Thus

$$\frac{\partial C_1}{\partial t} - D_x \frac{\partial^2 C_1}{\partial x^2} = 0$$

$$\frac{\partial C_2}{\partial t} - D_y \frac{\partial^2 C_2}{\partial y^2} = 0$$

$$\frac{\partial C_3}{\partial t} - D_z \frac{\partial^2 C_3}{\partial y^2} = 0$$

As obtained already the equations above can be written as,

$$C_1(x,t) = \left(\frac{\alpha}{\sqrt{4\pi D_x t}}\right) e^{-\left(\frac{x^2}{4D_x t}\right)}$$

$$C_2(x,t) = \left(\frac{\beta}{\sqrt{4\pi D_y t}}\right) e^{-\left(\frac{x^2}{4D_y t}\right)}$$

$$C_3(z,t) = \left(\frac{\gamma}{\sqrt{4\pi D_z t}}\right) e^{-\left(\frac{z^2}{4D_z t}\right)}$$

Putting together the equations above, we have

$$C(x,y,z,t) = \left(\frac{\alpha}{\sqrt{4\pi D_x t}}\right) e^{-\left(\frac{x^2}{4D_x t}\right)} \times \left(\frac{\beta}{\sqrt{4\pi D_y t}}\right) e^{-\left(\frac{y^2}{4D_y t}\right)} \times \left(\frac{\gamma}{\sqrt{4\pi D_z t}}\right) e^{-\left(\frac{z^2}{4D_z t}\right)}$$

$$\Rightarrow C(x,y,z,t) = \left(\frac{\alpha\beta\gamma}{(4\pi t)^{3/2}\sqrt{D_x D_y D_z}}\right) e^{-\left(\frac{x^2}{4D_x t}\right)} \times e^{-\left(\frac{y^2}{4D_y t}\right)} \times e^{-\left(\frac{z^2}{4D_z t}\right)}$$

$$\Rightarrow C(x,y,z,t) = \left(\frac{x}{(4\pi t)^{3/2}\sqrt{D_x D_y D_z}}\right) e^{-\left(\frac{x^2}{4D_x t} + \frac{y^2}{4D_y t} + \frac{z^2}{4D_z t}\right)}$$

Where, x = αβγ

The triple integration of C (x, y, z, t) from $-\infty$ to $+\infty$ is equal to the total mass injected at t = 0 at (0,0,0).

$$\int_{-\infty}^{+\infty}\int_{-\infty}^{+\infty} C(x,y,z,t)\,dxdydz = M$$

Thus

$$\int_{-\infty}^{+\infty}\int_{-\infty}^{+\infty}\int_{-\infty}^{+\infty} \left(\frac{x}{(4\pi t)^{3/2}\sqrt{D_x D_y D_z}}\right) e^{-\left(\frac{x^2}{4D_x t} + \frac{y^2}{4D_y t} + \frac{z^2}{4D_z t}\right)}\,dxdydz = M$$

$$\Rightarrow \left(\frac{x}{\left(4\pi t\right)^{3/2} \sqrt{D_x D_y D_z}} \right) \int_{-\infty}^{+\infty} \int_{-\infty}^{+\infty} \int_{-\infty}^{+\infty} e^{-\left(\frac{x^2}{4D_x t} + \frac{y^2}{4D_y t} + \frac{z^2}{4D_z t} \right)} dxdydz = M$$

$Considering, \ a = \dfrac{1}{4D_x t}, b = \dfrac{1}{4D_y t}$ and $c = \dfrac{1}{4D_z t}$, we have

$$\left(\frac{x}{\left(4\pi t\right)^{3/2} \sqrt{D_x D_y D_z}} \right) \int_{-\infty}^{+\infty} \int_{-\infty}^{+\infty} \int_{-\infty}^{+\infty} e^{-ax^2} e^{-by^2} e^{-cz^2} dxdydz = M$$

$$\Rightarrow \left(\frac{x}{\left(4\pi t\right)^{3/2} \sqrt{D_x D_y D_z}} \right) \sqrt{\frac{\pi}{a}} \sqrt{\frac{\pi}{b}} \sqrt{\frac{\pi}{c}} = M$$

Putting the value of the equations above, we have

$$\left(\frac{x}{\left(4\pi t\right)^{3/2} \sqrt{D_x D_y D_z}} \right) \sqrt{4D_x t\pi} \sqrt{4D_y t\pi} \sqrt{4D_z t\pi} = M$$

$$\Rightarrow \left(\frac{x}{\left(4\pi t\right)^{3/2} \sqrt{D_x D_y D_z}} \right) \left(4\pi t\right)^{3/2} \sqrt{D_x D_y D_z} = M$$

$$\Rightarrow \chi = M$$

Thus,

$$C\left(x, y, z, t\right) = \left(\frac{M}{\left(4\pi t\right)^{3/2} \sqrt{D_x D_y D_z}} \right) e^{-\left(\frac{x^2}{4D_x t} + \frac{y^2}{4D_y t} + \frac{z^2}{4D_z t} \right)}$$

$$C\left(x, y, z, t\right) = \left(\frac{M}{\left(4\pi t\right)^{3/2} \sqrt{D_x D_y D_z}} \right) e^{-\frac{1}{4t}\left(\frac{x^2}{D_x} + \frac{y^2}{D_y} + \frac{z^2}{D_z} \right)}$$

This is the solution of the three dimensional diffusion equation for the case of instanta-neous loading at x = 0, y = 0, z = 0 and at t = 0. If the release of mass is at any arbitrary location x= x_1, y = y_1, z = z_1 and at arbitrary time t = t_1 as shown in the figure below.

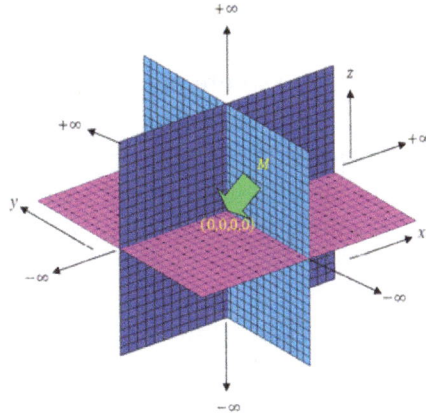

Loading of solute of mass M at (x_1, y_1, z_1, t_1) in a 3D unbounded domain

Then,

$$
\left.\begin{array}{l}
x' = x - x_1 \\
y' = y - y_1 \\
z' = z - z_1 \\
t' = t - t_1
\end{array}\right\}
$$

Thus the equation above becomes

$$
C(x,y,z,t) = \left(\frac{M}{\left(4\pi(t-t_1)\right)^{3/2} \sqrt{D_x D_y D_z}} \right) e^{-\frac{1}{4t}\left(\frac{(x-x_1)^2}{D_x} + \frac{(y-y_1)^2}{D_y} + \frac{(z-z_1)^2}{D_z} \right)}
$$

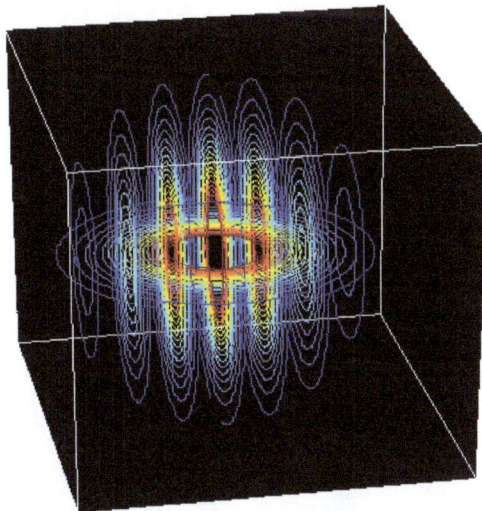

Contour plot at different vertical sections and one horizontal section shows the spatial distribution of solute at particular time caused by the process of diffusion

The equation above gives the spatial and temporal distribution of solute spread by the process of diffusion in a three dimensional unbounded domain. In this case, the visualization of the result is difficult as the solute plume will move in a three dimensional space. However, we can represent the plume using the contour lines at different vertical and horizontal sections at a particular time. Figure above shows spatial distribution of solute in the form of contours at different vertical sections at a particular time due to the instantaneous loading of solute at x = 0, y = 0, z = 0 and t = 0. This solution is obtained using equation.

Figure below shows the vertical and horizontal sections of the plume at a particular time t.

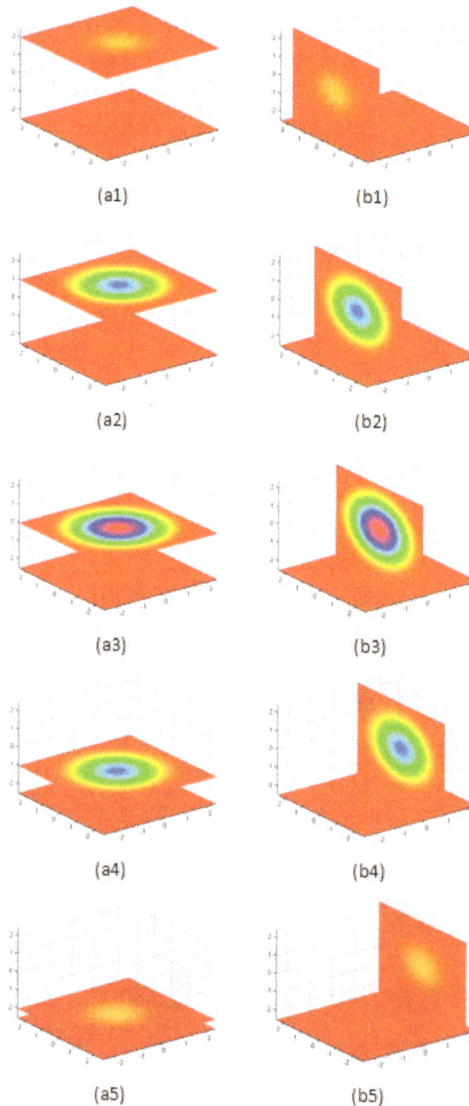

Spatial and distribution of solute spread by the process of diffusion at particular time t.
(a1) to (a5) Horizontal sections. (b1) to (b5) Vertical sections

References

- "Lack of data on fracking spills leaves researchers in the dark on water contamination". StateImpact Pennsylvania. Retrieved 2016-05-09

- Singh, B; Singh, Y; Sekhon, GS (1995). "Fertilizer-N use efficiency and nitrate pollution of groundwater in developing countries". Journal of Contaminant Hydrology. 20 (3–4): 167–184. doi:10.1016/0169-7722(95)00067-4

- Dean, R.G.; Dalrymple, R.A. (1991), Water wave mechanics for engineers and scientists, Advanced Series on Ocean Engineering, 2, World Scientific, Singapore, ISBN 978-981-02-0420-4, OCLC 22907242

- United Nations Environment Programme (UNEP) (2015). "Good Practices for Regulating Wastewater Treatment" (PDF). Retrieved 19 March 2017

- Suthar, S; Bishnoi, P; Singh, S; et al. (2009). "Nitrate contamination in groundwater of some rural areas of Rajasthan, India". Journal of Hazardous Materials. 171 (1–3): 189–199. doi:10.1016/j.jhazmat.2009.05.111

- Dingemans, M.W. (1997), Water wave propagation over uneven bottoms, Advanced Series on Ocean Engineering, 13, World Scientific, Singapore, ISBN 981-02-0427-2, OCLC 36126836 , 2 Parts, 967 pages

- ATSDR (US Agency for Toxic Substance & Disease Registry) (2008). "Follow-up Health Consultation: Anniston Army Depot." (PDF). Retrieved 18 March 2017

- Takahisa Okino, Mathematical Physics in Diffusion Problems, Journal of Modern Physics 6: 2109-2144 (2015)

- Lamb, H. (1994), Hydrodynamics (6th ed.), Cambridge University Press, ISBN 978-0-521-45868-9, OCLC 30070401 Originally published in 1879, the 6th extended edition appeared first in 1932

- State Coalition for Remediation of Drycleaners (2011). "A Citizen's Guide to Drycleaner Cleanup" (PDF). Retrieved 20 March 2017

- Landau, L.D.; Lifshitz, E.M. (1987), Fluid Mechanics, Course of theoretical physics, 6 (2nd ed.), Pergamon Press, ISBN 0-08-033932-8

- "Texas fracking site that spilled 42,000 gallons of fluid into residential area hopes to reopen". RT International. Retrieved 2016-05-07

- Phillips, O.M. (1977), The dynamics of the upper ocean (2nd ed.), Cambridge University Press, ISBN 0-521-29801-6, OCLC 7319931

- World Health Organization (WHO) (2006). "Protecting Groundwater for Health - Understanding the drinking-water catchment" (PDF). Retrieved 20 March 2017

- World Health Organization (WHO) (2011). "Guidelines for Drinking-water Quality" (PDF). Retrieved 18 March 2017

Permissions

Index

www.ingramcontent.com/pod-product-compliance
Lightning Source LLC
Chambersburg PA
CBHW061951190326
41458CB00009B/2841